BRASS FACTS ....

by

WARREN BRIGGS

Sterling House

Pittsburgh, PA

ISBN 1-56315-148-0

Trade Paperback
© Copyright 1999 Warren Briggs
All rights reserved
First Printing—2000
Library of Congress #98-88553

Request for information should be addressed to:

SterlingHouse Publisher, Inc.
The Sterling Building
440 Friday Road
Pittsburgh, PA 15209
www.sterlinghousepublisher.com

Cover design: Michelle Vennare - SterlingHouse Publisher
Typesetting: Drawing Board Studios

Printed in Canada

To our son Warren Briggs, Jr., whom we lost at too young an age. He was a poet at heart, a free spirit with a great sense of humor. He could laugh at life and the world with great abandon. His passing was a loss to us and the whole planet.

Table of Contents

Acknowledgments

"Think of what would happen to us in America if there were no humorists: life would be one long Congressional Record."

Tom Mason

We have always wanted to write a book just to keep up with our friends who have already done it. So here it is, a short collection of stuff written over the years.

There is no great wisdom here, mostly fun. There are some personal opinions. As movie mogul Samuel Goldwyn once said, "When I want your opinion, I'll give it to you."

In a telegram from the editor of the old *New York World* to his Washington correspondent, the editor wired, "Send all the details, never mind the facts." We have supplied mostly details here. A State Department spokesman once admitted, "Some of the facts are true, some are distorted, and some are untrue." We even invented some characters whenever necessary to make a point. A campaign aide to Senator Barry Goldwater admonished us, "Don't quote what he said. Say what he means."

We are indebted to Mark Twain, Will Rogers, Mike Royko, Yogi Berra, and other humorists for inspiration. They have made the world a funnier, better place to live.

We acknowledge borrowing some one-liners from the Petras' hilarious book, *The 776 Stupidest Things Ever Said* (Doubleday, 1993). We also borrowed a few from *Great Quotations* (used with permission and published by Successories, Inc., all rights reserved). The rest we have been hoarding for years. We also borrowed several incidents from *Will Rogers: Reflections and Observations* by Bryan B. and Frances N. Sterling, Crown Trade Paperbacks, 1986.

We wish to thank governments for screwing up, politicians for doing stupid things, and many others for being dumb, crazy, or funny. They are the fodder of comedy.

We have found over the years it's better to laugh than to cry. It hurts too much to cry.

We have had a lot of fun laughing. We hope you do too.

Thanks.
Warren Briggs

P.S. We also hope, as Yogi Berra once put it, "I really didn't say everything I said."

Foreword

As Henry Ford so aptly put it, "History is just one damned thing after another."

Brass Facts is a light-hearted look at the news events of the day, both past and present. Some of these are the momentous "big-ticket" items that will go down in history for posterity.

We fear, however, many of the lesser, unsensational, teensy-weensy but nevertheless important human events will be lost from sight and gone forever.

For example:

Deceased voters
Bureaucratic idiots
Cows on waterbeds

Navy Tailhookers
22 minute church services
Bungee jumpers

Old-fashioned political rallies
Squirrel attacks
Stolen garbage trucks
Outhouse history

Exploding commodes
and on and on and on.

We feel it our patriotic and historic duty to resurrect and perpetuate these atrocious events for evermore. Lest we forget, time marches on.

Some of these historic boo-boos include local color from Pensacola, Florida, but our local color will fit most localities. People and governments get into the same messes and do the same damned-fool things, no matter where they are located.

The philosopher Santayana said it for all time: "Those who fail to remember their history are condemned to repeat it."

WE HAVE READ YOUR PROPOSAL . . .

. . . AND ARE GIVING IT

SERIOUS CONSIDERATION.

ON BUSINESS

I would mind my own business if the government would give it back to me
Unknown

The meek shall inherit the earth but not the mineral rights.
J. Paul Getty

An oil well is a hole in the ground owned by a bunch of liars.
Mark Twain

The salesman got two orders: Get out and stay out!
Unknown

Economics is the only profession where a man can be an expert
without ever being right.
George Meany

Executives: People with their names printed on letterheads because
nobody can read their signatures.
Anonymous

A recession is when your neighbor loses his job. A depression is when you
lose your job. A panic is when your wife loses her job.
Unknown

The Joke's On Us

Lord, stir up strife amongst thy people, lest thy servant perish.
Invocation by struggling young lawyer (as told by Senator Sam Irvin)

While celebrating law week, the president of the American Bar Association (lawyers, not pubs) pleaded on national television for the American public to stop making fun of lawyers. "There are entirely too many lawyer jokes," he said.

He's right. There are a lot of marvelous lawyer jokes.

When former U.S. Attorney General Ed Meese visited Pensacola, he told the one about the lawyer who arrived at the Pearly Gates to find bands playing, speeches being made, and a great celebration in progress.

"What's going on?" he asked innocently.

"We're celebrating in your honor," said St. Peter. "You're the only person we have ever had here who was 193 years old."

"I'm not 193," protested the lawyer. "I'm only 81."

"That can't possibly be true," replied St. Peter. "You *must* be 193. We verified your age based on the time you billed your clients."

This is funny unless you have ever been billed by a lawyer. There are other great lawyer stories. There's the one about the lawyer who was questioning his client on the stand during a trial:

"Are you the defendant in this case?" asked the lawyer.

"No, suh," replied the client. "I'm da one who stole da chicken."

And then there was the lawyer who bribed one of the jurors in a murder trial to try to get his client off with life instead of the death penalty.

The jury came back with life. The lawyer was overjoyed, congratulated the juror he had bribed, and asked him how he did it.

"Well, it wasn't easy," said the juror. "The first vote we took was 11-1 for acquittal."

So there you have some lawyer jokes. There are a million more.

But this is serious business, folks. You may need a lawyer some day, and that's no laughing matter. Everybody is suing everybody else these days at the drop of a hat, or even less. We are a sue-happy society. We have more lawyers per acre in America than all the rest of the world combined.

Consider the case of the homeless couple who were recently run over by a train while making love on the railroad tracks. This is an honest-to-God true story which was reported in the press.

Naturally, the couple is now suing the railroad for interfering with their connubial bliss. Obviously the railroad will also need a lawyer (or several hundred of them) to defend itself against the couple's lawyer, who is not homeless.

Then there was the driver who drove his car through a phone booth which was occupied at the time. Never mind the poor guy in the phone booth. The driver is suing the phone company for putting the phone in the wrong place.

That's the way it goes. So let's get serious and knock off all these dumb jokes about lawyers. You never know when you may need one.

Take Two Aspirin

> I'm often wrong but never in doubt.
>
> *French Surgeon*

Lawyers did not think our lawyer jokes were very funny and have now demanded equal time for doctor jokes.

Lawyers don't generally care for doctors and *vice versa*. Lawyers define a doctor as a person who still has his tonsils, adenoids, and appendix.

Doctors accuse lawyers of running up exorbitant malpractice costs, and lawyers claim doctors and hospitals charge too much and make too many mistakes.

Lawyers love doctor jokes.

They tell the one about the doctor who complained to a lady patient that her check came back. Replied the lady, "So did my arthritis."

A child swallowed a shotgun shell. The parents panicked and called their doctor. He advised, "Keep him quiet, and for God's sake, whatever you do, don't point him at anybody."

After a counseling session, a psychiatrist told his patient that his problem was he hated his brother. "But I don't have a brother," protested the patient.

Replied the psychiatrist, "Look here, young man, if you don't cooperate, I can't help you."

Another psychiatrist told his patient, "You're crazy."

"I want a second opinion," demanded the outraged patient.

Responded the psychiatrist, "Well, all right, you're also ugly."

A friend had the shakes so the doctor prescribed a little wine. The friend got so steady he couldn't move.

A doctor informed his patient his temperature was 102 degrees, "but it's only 99.1 with the windchill factor."

Doctors' offices are filled with piles of ancient printed materials.

Testified one patient, "My doctor was amazed to discover I had a disease that hadn't been around for almost a century. Turned out I caught it from a magazine in his waiting room."

Doctors don't always have time to explain treatments fully. One patient looked unhappy so the doctor asked him what the trouble was. "Plenty! On account of my arthritis you told me to avoid all dampness. You've no idea how silly I feel sitting in an empty bathtub and going over myself with a vacuum cleaner."

Finally, physicians are born with notoriously bad penmanship. There is the story of the lady who sent out invitations to dinner and received an absolutely illegible letter of reply from a doctor. She couldn't read it.

"I have to know if he accepts or refuses," she declared.

"If I were you," suggested her husband, "I'd take the letter to the druggist. Druggists can always read doctors' writing no matter how bad it is."

So she went to the druggist and handed him the letter. He went to the back of the store and came back a few minutes later with a bottle.

"Here's your prescription, madam," he said. "That will be $27.40."

The lawyers are even.

Sawmill Days

> To err is human; to forgive is not company policy.
> *Great Quotations, Inc.*

We recently renewed old sawmill ties at Sawmill Day in Century, Florida. Century is the last best vestige of a southern sawmill town and has been designated an historic district. The sawmill was built in 1900 — hence Century — and at one time was the largest sawmill east of the Mississippi River.

On Sawmill Day sawmill folks and relatives and friends gather from miles around. The population swells ten times. Sawmilling is a brotherhood, a family affair. There was great food and fun and excitement at this last Sawmill Day. We had a big parade and lots of politicians. We even had a runaway horse (which was not in the program).

Now a grave injustice to sawmills has been perpetrated. A Mr. Dave Goodwin, writing in the *Pensacola News Journal,* referred to sawmills as boring. *Boring*! Anybody who calls a sawmill boring has never been within 100 miles of one. This is an insult. Mr. Goodwin must think sawmillers just sit and watch the trees grow. Let us enlighten him.

Sawmilling is an exciting and dangerous sport. It is an ancient and honorable profession. Sawmills built the world, at least most of it.

It all begins in the logging woods with the loggers who are first cousins to sawmillers, kindred spirits. Oxen or Caterpillar tractors plow belly-deep through the swamps dragging out the logs. He-men loggers bearing heavy crosscut or chain saws wade through the muck. They cut one tree, and then hundreds more are planted in its place so Mr. Goodwin can sit and watch them grow.

A tree is a graceful thing standing and swaying in the wind. Cut and falling, it becomes a lethal weapon, tons of danger. Men die crushed by falling trees. Being squashed by a felled tree is considered a badge of honor by loggers. It is a very permanent honor.

Then the log arrives at the sawmill ready for action. Saws slash away. Sawdust flies; the smell of it gets in your nostrils and in your blood. Fresh and clean. The log rides back and forth through the head saw on a steam-driven "carriage" much like a catapult on an aircraft carrier. Every once in a while there is an explosion, and the carriage flies out the end of the sawmill with the blocksetter riding on it. Sometimes he survives. Or the band saw breaks and wraps itself around the sawyer. He had better duck in time.

No self-respecting, seasoned sawmiller has all his fingers. They have been sacrificed to the profession. To order two short beers, a real sawmiller holds up both thumbs because all the rest are missing.

The sawyer slices slabs from the log as it races back and forth in front of him. Edgers then shape the slabs into lumber. Edging strips are sometimes picked up by the edger saws and hurled at the speed of an arrow. They can go right through a man and sometimes do. He had better not blink for any extended period of time.

There is no waste in a sawmill. The bark, peeled off by giant scrapers, goes to the fuel house to feed the boilers. So does the sawdust. A whining chipper grinds all the leftover, unuseable slabs and strips into small chips to make paper. The profit is in the chips. Occasionally a man gets sucked into the chipper. A man who has been through a chipper will also fit through a strainer.

There is a sophisticated wood preservation plant using a large, long cylinder which looks much like a space rocket. It's full of propane gas. Light a match at one end and you had better hold on tight because you're on your way to the moon.

Back in the office there is the thrill of the "hunt" for markets. Bowling alley floors ship to Japan and Ireland and Germany. Italian prime goes to the Mediterranean. Timbers go to the South African gold mines. The Caribbean and Bermuda take framing lumber. The foreign merchants come to visit and watch and eat at the boarding house and bargain and buy.

The sawmill boarding house is an institution. It goes with the business. It offers bed, board, fun, friendship, tall tales, midnight trysts, intrigue, and the best cornbread muffins this side of heaven.

The mill runs day and night in good times. Then you work all weekend to plug the leaks and patch the machines and grease up and heal your wounds so you can run again on Monday.

The night shift winds up at a late hour, and the sawmill crew heads for home in the dark. The shotgun houses, lined up in a row, all look alike — duplicates. You have to count down from the boarding house to find your own. One worker miscounts and goes to bed in the wrong house. After this happens several times, suspicion grows. Shots are fired. Sawmilling is dangerous work even after hours.

Every good sawmill man deserves a good fire occasionally. One owner in Alabama had a fire which was covered by insurance. When the insurance company then offered him hurricane insurance he said, "Sure, but how do you start a hurricane?"

The sawmill in Century burned in 1959. Careless welders started the fire. Fire in a sawmill is a constant dread. You don't smoke anywhere in or around the mill. But as the fire roared and the roof fell in during the 1959 fire, the finance officer, who was watching the fire, lit up a cigarette. "I guess it's OK to smoke now," he said.

A sawmill is a series of crises and explosions. It is a very special place for a very special breed of people. There are no sissies in sawmills. There is a special place reserved in heaven for sawmillers in a sawmill which runs perpetually without grease. No breakdowns; lumber pouring out 24 hours a day forever.

Boring? Hmmft!

The Gainers

> Buy stock when it's low, and sell it when it's high.
> If it doesn't go up, don't buy it.
> *Will Rogers*

Garden Street is Pensacola's Wall Street. A branch of our Wall Street also runs north on Palafox Street to Dean Witter's brokerage office where monthly a stock investment club of 21 ladies meets to plot, scheme, and invest. This is the infamous 21 Club.

These ladies are very successful investors, but they and the whole world should know that they owe their success to a pioneer group of men who formed America's very first investment club in the early 1900's in Century, Florida. This group was named "The Gainers."

We have dug up the history of the Gainers and are able to reprint a copy of their 1963 annual report. Not even the names have been changed to protect the innocent.

The following annual report should be read in candlelight to the accompaniment of soft music (preferably violin and harp):

"The Gainers, who have now lost only $117.42, met at the Alger office on Monday.

"The audit committee arrived at 6:30 and managed to break into the building by 6:45. They were unable to find the books, whereupon the treasurer was summoned. The treasurer refused to surrender the books without proper credentials and subpoenas. The audit committee obtained the books at 7:29.

"At 7:30, the president called the meeting to order. After some slight delays, the secretary found the minutes from the preceding year. They were approved as read.

"Mr. Carl Jones of the audit committee reported that as far as they could tell, the books were neatly kept. The president complimented the audit committee on this nice report.

"The treasurer rendered the financial report amid ringing cheers from stockholders and audience. Copies are attached for those who missed the meeting.

"The president then called upon the nominating committee for its report. After some fumbling and bumbling, they retired to the men's room to caucus. There was a long, embarrassing silence.

"Mr. McNeel began audibly fingering his proxies, whereupon Mr. M. C. Leach, a minority stockholder, rose to the occasion. Mr. Leach recited the pitiful history of the market in 1962 and announced he was ready to forget the past and let bygones be bygones. (The president began to weep silently in the background.)

"Mr. Leach, dabbing his own eyes repeatedly with a moist *Wall Street Journal,* continued. He felt that the new management had not had time to average out its losses, he said. Then, in a magnanimous act of unequaled charity, he nominated the entire management for another term. Mr. Jones, another minority stockholder, seconded; and the vote was unanimous. Dave Turpin was elected a new director.

"The president, now weeping uncontrollably, was unable to speak further. The meeting was adjourned voluntarily. Several members helped the president out of the building. Later he sent his word of appreciation to the minority stockholders for their fine representation and support at the meeting."

The moral of this story is probably that what goes up can also fall down. The "Gainers," renamed the "Losers," have since disappeared along with the treasurer, the books, and the audit committee.

Amen!

Windstorms

> One person out of eight has an accident;
> the other seven have accident insurance.
> *Anonymous*

We just received a "Dear John" letter from our insurance company: "Your homeowner's policy is canceled, period."

No regrets, no thanks, no explanation, no nothing after 30 years of paying exorbitant premiums to these pirates. We were irate so we wrote the president of the company expressing our rage at being unceremoniously dumped. We were not polite.

We received a nice letter from some underling expressing regrets and explaining that the company had to reduce its windstorm exposure in Florida. We still had no insurance; we were madder than ever. Since we were not pacified, we called this guy up. "Why us?" we shouted.

"You sound angry," he said, which was a slight understatement while we were actually frothing at the mouth.

"You darned betcha we're mad. How would you like to be cut off with an insulting announcement like this?"

"I wouldn't like it."

"What am I talking to you for anyway? I wrote the president. Did he die?"

"The president is very busy."

"I guess he's too busy to care what kind of lousy customer relations he has. He's not user-friendly."

"Well, why don't you talk to your local agent?"

"Because he didn't cancel our policy; you b _ _ _ _ _ ds did. Where do I get coverage now?"

"I don't know." We couldn't stand anymore of this so we slammed down the phone which broke into several pieces.

We wrote the president again. This time we got a personal call from a slightly higher peasant.

"Can I help you?" he asked cautiously. He obviously had a report on our earlier messages.

"Who are you, and where's your president? Does he work there?"

"He got your message."

"Does he know about these stinking cancellations?"

"The government approved the form."

This idiocy brought on severe apoplexy. When we recovered, we shouted, "Can't you do any better than the government? Our house went thru two hurricanes down here with no damage. The only wind damage we've had comes from your office."

"We're sorry."

"Sorry doesn't cut it. Where do we get insurance now? You b _ _ _ _ _ ds have cut us off."

"I've reviewed your policy, and we're going to renew it."

They say "Don't get mad, get even" so we decided to get even. "We wouldn't renew with your sorry company for all the tea in China. We'll take our business to somebody who appreciates it. Thanks anyway."

We hung up the pieces of the phone. Then we suddenly realized we had just canceled our own insurance. Our agent had to call and eat our crow.

This story is true. Only the names have been omitted to protect the guilty.

God Bless America and the Lusty Lady

> When a great many people are out of work, unemployment results.
> *President Calvin Coolidge*

The labor unions played a big part in the 1996 Democratic Convention. One-third of the delegates were union members even though in the U.S. union membership has fallen from 35% of workers in 1954 to 15% in 1995.

The unions endorsed Clinton early on and dunned members $35 million to finance an ad campaign to beat Dole.

The largest and strongest union is the national teachers union, the National Education Association. Teachers have joined other government employees, hod carriers, sewer workers, plumbers, welders, truck drivers and other laborers in the union vineyard. The NEA generally opposes charter schools and other innovations in education except more money.

As serious as they are about organizing more workers and beating Dole, at least the unions haven't lost their sense of humor. The Associated Press recently reported two union stories worth passing along.

The nude dancers at the Lusty Lady Bar and Grill and Ballet voted 57-15 to join Local 790 of the Service Employees International Union. It seems the nice ladies, advertised as "nasty nymphs," were dissatisfied with working conditions, including pay. Furthermore, it seems that some of their customers had been doing some unauthorized peeping and pinching on the fair damsels.

The dancers were ecstatic over their new condition of servitude. One naughty nude ballerina named Amnesia spoke for the group. "There ain't gonna be no more peeping and pinching with the union in here. Management's gotta shape up along with our clients. I'm black and blue all over," she said while blushing. Score one for the union.

ASCAP scored number two for the unions. ASCAP is the American Society of Composers, Authors, and Publishers which protects the copyrights on all music in the U.S. So guess what? ASCAP gets royalty money when you borrow one of its tunes.

Now, it seems that ASCAP doesn't even want you wandering around town humming any of their copyrights unless you ante up. You don't sing *Happy Birthday* to your mother-in-law without paying. You don't sit around your campfire toasting marshmallows and harmonizing on *Auld Lang Syne* without mailing in a check to ASCAP.

How could this happen in a free country? The American Camping Association is up in arms. Campgrounds would pay royalty fees from $77 to $257 just to let their happy campers sing in the showers.

We couldn't believe this news ourselves so we went to visit C. Sharp Major, an ASCAP composer. "Are you guys serious?" we asked.

"You damn betcha we're serious," he shouted. "It took a lot of inspiration to write my latest masterpiece, *Rappa Clappa Trappa*. You think I want every 2-bit yodeler in the country singing my tunes free-of-charge? No way, you gotta pay."

"How will you collect?" we asked.

"That's not my job. Go see our lawyer." So we did. He was ASCAP attorney B. Flat Minor, Esquire. He was counting quarters.

"How on earth are you going to keep up with collecting all these royalties?" we wondered.

"That's easy; nothing to it. We have spies all over. We have members standing on street corners with tin cups. You sing our songs, you pay. Drop a quarter in the cup."

"That won't catch everybody," we said.

"That's just the beginning," said lawyer Minor. "We have listening devices on poles. We install meters in your house and car. You can't escape us. When you feel the urge to sing a little opera in the bathtub, we can hear you. You'd better stick a quarter in the meter. We catch you cheating, you're in deep doodoo."

'What's the penalty?" we asked.

"Listening to Tiny Tim records," he warned.

As we left, Minor was still counting quarters. If you feel the urge to whistle a tune, be careful. Have your quarter ready. The ASCAP big brothers are watching and waiting for their cut. We hummed to ourselves, "God Bless America." Oops! There goes another quarter.

Crash

> I'm opposed to millionaires, but it would be a mistake to offer me the job.
> *Mark Twain*

And away we go. Down! The stock market bubble finally popped.

First Hong Kong crashed, then American investors decided they had better get out before the Hong Kong "flu" bug infected our markets. Wall Street went down several hundred points, then bounced back up several hundred like a yo-yo.

Over the weekend investors started thinking about all the Asian doom and gloom. "Oh my God," they decided, "we'd better bail out while we still have a bucket."

On Monday, October 27, 1997, remembering the 1987 market debacle, investors rushed to pay phones, cellular phones, faxes, computers, brokers' offices, or however they could get their hands on a stock salesman and pulled the plug. Down she plunged, 554 points, a world's record. Nothing could stop it including the "circuit breakers." The market is psychological: when you gotta go, you gotta go.

Then Hong Kong recovered for a day. Guess what? Back up we went 337 points on Tuesday, another world's record; 1.2 billion shares were bought and sold, still another record. Now we're teetering on the brink; now we're up, now we're down, like the bouncing ball waiting on Dr. Alan Greenspan to blink or burp or give some other distinguishable sign up or down.

What gives here? The investors giveth, and the stock brokers taketh away. Right or wrong, they always win. They get your money going in, and they get your money coming out.

Before the crash their advice was "a strong market and economy still rising. Buy!" After the crash on Monday the advice was "a minor correction in a strong market and economy. Great buying opportunities. Buy before it's too late." They gotcha. It's always a time to buy.

So what do we investors do now? Buy? Sell? Hold? Nobody seems to know except the all-knowing brokers so we spoke to Seldum Wright, a broker with Foolem,

Fleecem, Stickem, and Soakem on the 50th floor of the Good Times Towers on Wall Street.

"Is the worst over?" we asked.

"Shhh! Don't ever use that term," warned Wright. "You'll scare everybody to death. Everything is lovely, just lovely."

"Is that so? Maybe you're too young to remember 1929. People in New York were jumping out of windows. The hotels put in mortuaries to handle the investors."

"Never happen now. But we're prepared just in case."

"How's that?"

"See that big plate glass window at the end of the hall. The sign says, 'In case of emergency, break glass!' If we have to go, that's the way out."

"Why is that bungee cord hanging down the front of the building?"

"That's for investors who want to get a feel for the market. They can bounce around for a while."

All this did not sound very reassuring so we sought out investor I. "Ben" Skrood to see how he made out.

"Are you optimistic or pessimistic?"

"I'm a pessimist. Wright sold me a bunch of junk on Friday, and it crashed on Monday. Gone. Kaput."

"You aren't thinking of jumping, are you?"

"Not yet. But I'm going to push Wright out."

"What's that on your back?"

"That's a parachute. If I go out, I want to be able to change my mind before I hit bottom. The damned market might go back up on my way down."

Pessimists are optimists with experience.

Robbing Hoods

> I regret that I have but one law firm to give to my country.
> *Adlai Stevenson*

Where is Robin Hood when we need him?

Governor Chiles and twelve expensive lawyers have just wrested 11.3 billion — yes, *billion* — dollars from the cigarette companies which went up in smoke in court.

Now it seems these twelve legal eagles need 25% of the take as their reward for surrendering the cigarette makers to the State. That amounts to roughly $250 million per lawyer, rounded off to the nearest 50 million. At $250 per hour, say, which should bring in some pretty sound legal advice, that would mean each lawyer put in 1,000,000 hours of work on this case. At 80 hours of legal work per week that amounts to 12,500 weeks or roughly 240 years of intensive legal effort.

One attorney said he spent $750,000 on the case; thus his profit would reduce to $249,250,000. The good old highwaymen of yesteryear never had such a good deal. What they wouldn't give (or take) for a job like this.

The lawyers are screaming, the governor is moaning, and mediators are trying to negotiate. Maybe they would compromise for 50 or 100 million and prove the lawyers are statesmen and patriots after all. Chiles says it's up to the cigarette companies to pay the lawyers. He wanted all the state's take to go to children's health. That's the way Robin Hood would do it.

We have scratched our heads to remember similar acts of piracy. We recall Bonnie and Clyde, Jean Lafitte, and the famous Willie Sutton. When asked why he robbed banks, Willie said, "Because that's where the money is." The cigarette companies certainly do have the money.

Then there was Jesse James and the James brothers. They pillaged the banks in the midwest, but their loot was paltry compared to $249,250,000. The James brothers did not make charitable contributions, distribute the wealth, or provide for the care and feeding of children. Eventually a posse filled them full of bullets. Their security was lax.

In the business of taking food from the mouths of babes, we were reminded of two recent examples of heinous acts. You may remember the great Girl Scout cooky heist. Two masked bandits held up a Girl Scout booth where the young ladies were plying their cooky trade. Looking down the muzzles of 6-shot repeaters, the scouts surrendered all their cash. The criminals also ate the cookies.

Another more recent case, even worse, involved the great lemonade robbery. While engaged in capitalistic free enterprise, a little girl was accosted by a bandit at her lemonade stand and forced to hand over all the profit and then some. To add insult to injury, the thief drank her lemonade and escaped red-handed. The police got no clues and no lemonade.

In all the annals of known crime, we have never seen such a rash of objectionable behavior. Crime may be down, but shameless acts are up. You just never know where the next attack will come from. We all need more security.

Someone said that the easy way to teach children the value of money is to borrow it from them. The other way is to steal it. Where, oh where is Robin Hood?

Caveat Emptor

> We're from the government, and we're here to help you.
> *The Government*

The government intends to save us from ourselves whether we like it or not. Reams of government paperwork, laws, and regulations protect us.

The current fight targets smoking. Congress and the states battle the cigarette companies who have apparently lied about addiction and connived to snare children at tender ages.

Our own father started smoking behind the barn at about age seven and kept right on until he quit cold turkey at age sixty. Then we caught him smoking again at eighty. We asked him why. "Well, at eighty," he said, "what the hell." He died still happily smoking at ninety-two. He didn't believe in too much government; he knew exactly what he was doing.

Drugs and liquor are also big consumer problems. From 1917-1933 we prohibited whiskey in the Constitution. The result was bootleggers, Al Capone, homebrew, and bathtubs full of gin. Prohibition didn't work; the buyers didn't buy it.

Current and tough laws prohibit drugs, but drugs pour across the Mexican border and sail into Florida. Gradeschoolers use pot. Users buy the stuff on most street corners. Whatever we're doing to stop it ain't working. The users could stop it if they stopped using.

Laws label every item we buy with some kind of warning or advice. Some manufacturers go to amusing lengths to comply:

A knife manufacturer warned buyers, "Keep out of children."

A shirt manufacturer advised, "Do not iron clothes while you are in them."

A food processor notified buyers, "Peanuts. Open package. Eat." That certainly made sense.

The government hands out lots of advice about diet and calories and cholesterol and exercise. We label food by contents, good or bad, no fat, low calorie, no cholesterol, sugar-free, etc. The government has removed all the dangers and all the taste.

The government experts frequently change their minds. Butter was out, margarine in. Now margarine is out, butter back in.

Eggs were bad so we took out the yolks and made scrambled whites. Yuck! And no bacon to go with them. Now we can sneak in a little yellow while we take our cholesterol pills. Fat was all bad; now we need some fat. Who knows?

The government tells us what to eat daily to survive. Here are just a few of the government's daily requirements: broccoli, cauliflower, spinach, sardines, bananas, herring, celery, wine, tomatoes, anti-oxidants, beta carotene, olive oil, soy beans, fish, milk, garlic, rye bread, selenium, and all the vitamins and minerals from A to zinc, along with real food to stay alive. You can hardly stuff all this down in one day.

There is now a run on the sex pill Viagra. Scientists also work on pills to stop wrinkles and to grow hair. They found a new hair tonic — it doesn't really grow hair, but it shrinks your scalp to fit the hair you have.

Add to this mix all the lawyers it takes to sue everybody for product liability. The U.S. has 800,000 lawyers, one-half the world's supply, twice as many per capita as Britain, four times as many as Germany, twenty times as many as Japan. Enough for every man, woman, and child.

So where do we consumers fit in? Do we victims have any responsibilities?

Buyers beware!

2 + 2 = 5

> To err is human; but if you really want to foul something up,
> give it to a computer.
> *Anonymous*

World chess champ Kasparov played a chess match with an IBM computer. The computer almost won. It did win the first few games until Kasparov caught on and got hot.

Reputedly the computer could analyze 10 billion moves in three minutes. How could any human beat that? Computers are becoming invincible — or are they?

It now appears that a terrible worldwide computer glitch will take place at 00:01 a.m. on January 1st in the year 2000. It seems that the geniuses who programmed all our computers used just 2 zeros (00) to represent the year 2000.

But it turns out that the computers aren't so smart after all; they can't tell whether 00 means the year 2000 or the year 1900. That certainly seems ridiculous; any adult dummy can figure that out. In fact our ten year old granddaughter won't have any trouble telling the difference between the year 1900 and the year 2000.

By the year 2000 computers will dominate your life. You won't be able to pay bills, shop, bank, order, get married, or be buried without a computer.

So what does all this mean? It means that all hell will break loose in 2000. Experts estimate that it will cost 600 billion dollars worldwide to fix the mess. That's $300 billion for the U.S., which owns half of the world's computers. That sum equals the savings-and-loan bailout which broke the Bush administration.

The problem will occur on Mr. Gore's watch assuming he's elected. It couldn't happen to a nicer guy. The results of this gaffe in the year 2000 are almost too horrible to consider. Here's what will happen:

President Gore signs the budget bill. The computer throws it out because it thinks it's 1900, and William McKinley was president in 1900. The computer wants McKinley to sign the bill.

You file your income tax return on your computer. The computer says no tax is due because the income tax wasn't invented until 1913. You fail to pay your tax. The IRS is not happy, and you go to jail.

You order airline tickets for your vacation. The computer takes your money but fails to make the reservation because even the computer knows that the Wright Brothers didn't fly until 1903. Your vacation goes up in smoke when you arrive at the airport with all your bags packed and no ticket and no place to go.

And so on. By now the computer is really beginning to enjoy screwing up the whole democratic system. There will be no end of tragedies unless the computer programs can be fixed.

In the meantime, how do we explain to our granddaughter what is going on? The difference between 1900 and 2000 is simple arithmetic taught in elementary school. Computers as dumb as that will never get into college, but they might be elected to public office.

Beware!

Fresh Dollars

Crooked Dough: Counterfeit and Pretzels.

Anonymous

The counterfeiters are pouring out bogus currency, and the federal government is trying to design new bills with all sorts of secret features camouflaged to thwart the crooks.

We went to visit treasury official Ian Dett to see how the program was working.

"Could you tell us how you disguise your new bills?" we asked.

Dett slammed his door shut and jerked down all the window shades. "What? Are you kidding? Good God, man — this is highly classified! Top, top, top secret! We can't tell every Tom, Dick, and Harry. For all I know, you may be one of them."

We assured Dett we were one of us, not one of them. "But what about the public? People need to know the difference."

"I can't tell you," said Dett. "Go ask the counterfeiters. Here's a list."

So we drove over to visit Juan Dolár, ace counterfeiter, who just happened to be running his press at the time. He had just finished a fresh batch of $100 bills which were hanging up on the clothesline to dry.

"How can we tell these are counterfeit?" we asked innocently.

"That's an insult," said Dolár. "You can't tell the difference. These are made to government specifications."

"Don't you feel guilty passing out these worthless bills to innocent victims?" we wanted to know.

"Are you kidding?" snorted Dolár as he rammed his press into high gear. "Mine are just as good as the government's. They print money and give it away all the time. I'm just helping the economy. "

"How so?"

"We're $5 trillion in debt. We don't have enough money to pay it off. I'm just helping out. It's the patriotic thing to do. If I circulate enough of this stuff, we'll soon be out of debt."

"Does the government know about this?"

"Naw, no need to tell them yet. When we get out of debt, Secretary Rubin will want to give me a medal. I'm bailing him out. I've got the answer to the budget problem, and we won't have to shut down the government or default on the debt. It won't even take seven years."

"But there's nothing behind your currency," we protested.

"I'm behind it. The government ain't backing up its debt. I'm doing it with my dollars. Washington won't even cut spending in order to balance the budget. They're the crooks. They're spending money they don't even have, but I have it."

"What about the ordinary guy on the street or on welfare?"

"No problem. I've got a plan to run these off on Xerox machines. They'll meet all government specs, and anybody can do it. This will end welfare as we know it!"

We are reminded of the story about famous Yankee baseball manager Casey Stengel who once asked baseball great Yogi Berra, "What would you do if you found a million dollars?" Said Yogi, "I'd find the fellow who lost it, and if he was poor, I'd return it."

Yogi and Dolár may be on to something important here. We will forward these ideas on to President Clinton. He'll like them.

Good luck!

Don't Ask, Don't Tell

> There is nothing wrong with having nothing to say—
> unless you insist on saying it.
> *Anonymous*

Going for a job interview?

According to the 1996 Florida Employment Law Manual, here is a list of things your interviewer is not supposed to ask you about:

Height and weight
Color of hair and eyes
Sex
Age
Plans to get pregnant
Pregnancy (if obvious)
Arrest record
Credit record
Legal history
Marital Status
Dependents
Father's name
Mother's name
Maiden name
Birthplace
Religion
Race or color
Memberships
Language
National origin
Name of minister, priest, or rabbi

These rules must make for a very interesting and enlightening interview. We wondered how such a modern interview would work out so we accompanied Olive Owtawurk during her search for employment.

Olive made an appointment for a job interview with the Acme Sanitary Sewer Service, Inc. Founder and president Oliver Overflow greeted us at the door. We sat down to a champagne and caviar snack.

"Welcome," said Overflow.

"Let's cut out the crap," said Olive. "What do you do here?"

"Everything in sewers."

"That's not good enough."

"We clean sewers," said Overflow who then went into great detail on how sewers work.

"Do you carry all the necessary insurance?" Olive wanted to know.

Overflow dumped out all his policies.

"Now," asked Olive, "what do you pay?"

The president was visibly embarrassed. "We only pay $12 an hour plus all benefits including wet suits, hip boots, and gas masks," he admitted.

"That's not enough."

"Well, what about $15?"

"You're getting warmer. What about vacations?"

"Anything you want," said Overflow. "You get all holidays off including Valentine's Day, your birthday, anniversary, and anything else you can think of. You get a month's paid vacation to start off and then a week or two every month."

"What are your working hours?"

"Whatever you want. You just work when you feel like it. When you don't feel like it, you take paid sick leave. We also take naps after lunch."

"What about my family?"

"They can all work here if they want. We send your kids to nursery school, private high school, and Harvard or Yale. Are you married?"

"You can't ask that."

"Oops, I forgot. Now, how about a job?"

"Well, I don't know yet," said Olive. "I'm still not convinced. I'll have to think about it."

"Oh, please, please," pleaded Overflow, getting down on his hands and knees. "What else can we do for you? Here, try some more champagne."

"That sounds like a bribe."

Overflow apologized, tears running down his face. "Oh, Lordy, forgive me. I forgot about the manual. We just want to help you and your family. Are you on welfare?"

"That does it," said Olive. "According to the manual, you're not allowed to ask that. This is a job interview. I ask the questions. You're in violation. I'm suing."

Overload was now prostrate on the floor, sobbing uncontrollably. "Please give me another chance," he begged.

Olive agreed to think it over and get back to Overflow. As we left, he was praying. Olive has since decided to sue.

We feel a lot better now that we know how a job interview is supposed to work. We hope this information will help you find a job without being asked a lot of unnecessary, embarrassing questions such as the color of your eyes and your sex.

ON CRIME

Hijackers should be given a rapid trial, according to due process,
at the airport, then hanged.
Los Angeles Police Chief, 1973

A gun is a recreational tool, like a golf club or a tennis racket.
You can kill someone with a golf club, you know.
Martel Lovelace, NRA Official

The streets are safe in Philadelphia; it's only the people
who make them unsafe.
Frank Rizzo, ex-Police Chief and Mayor

The more killing and homicides you have, the more havoc it prevents.
Chicago Mayor Richard Daley

If a person is innocent of a crime, then he is not a suspect.
Former Attorney General Edwin Meese

Bang, You're Dead

Outside of the killings, Washington has one of the lowest crime
rates in the country.
Marion Barry, Mayor of Washington, D.C.

The following statistics show the number of people killed by handguns in 1992:

22	Great Britain
13	Sweden
91	Switzerland
87	Japan
10	Australia
68	Canada
10,567	U.S.

We don't know how many of these shooters were bad guys wearing black hats who needed to be shot. The total for the U.S. in 1993 was approximately 12,400, a "modest" increase.

Which naturally brings up the issue of gun control; this is a very touchy subject.

Certain weapons are already controlled. As a private citizen you can't own and operate a tank, a 105 howitzer, a 50-caliber machine gun, a bazooka, or a rocket launcher. This is already gun control. It's here. You can't use heavy artillery in spite of the Constitution.

A current argument rages over certain semiautomatic assault weapons which fire lots of bullets in a hurry. You can squeeze out 20-30 rounds in a matter of seconds, almost like a machine gun. What to do about these spray guns?

Several hunting groups recently called for a strict limit. These hunters said they didn't need more than 4 or 5 shots at one time. "You don't need to kill the whole herd with one blast," they said. This makes some sense.

The real rage is over handguns. We now have a waiting period. Unfortunately, criminals do not stand in line waiting for guns. They steal them or buy them on the black market or shoot the victim with his own gun.

There is a new proposal to keep handguns out of the hands of kids. This makes sense. Schools are arsenals. They have become shooting galleries with students armed to the teeth. Children have to go through metal detectors and wear bulletproof underwear.

To get the public's reaction to these problems, we approached citizen Sam S. "2-Gun" Slinger who was practicing quick draws in his backyard. He was wearing a white hat and twin 6-shooters, one on each hip. As we neared, he shot a milk bottle off a fence post with his first round.

"What about gun control?" we asked.

"Depends," he said.

"How much gun is enough?" we asked.

"Well, let me put it this way," said 2-Gun. "In the good old days we didn't fight duels with submachine guns. It ain't gentlemanly."

A quick round from his other six-shooter conked a stray possum who was innocently passing by.

"So what's the answer?" we asked as we ducked.

"I'm a single-shot man, myself," said 2-Gun. "All you need is one good shot per burglar. That's the decent, honorable thing to do. You don't need to fill them up with lead. They get too heavy to carry."

"What if you miss?" we asked.

"Don't," he said.

"What about hunting?"

2-Gun thought a minute. "You don't need a 155 howitzer to shoot a deer. A direct hit and you'd have to scrape the meat out of the trees."

He fired off one more practice round for effect. It went through a bothersome neighbor. He apologized and said, "You can't shoot ducks with antiaircraft guns. There wouldn't be enough left for gravy." We left him twirling his 6-guns like Quick Draw McGraw.

We don't know how all of this will end up. Politically, it's a free-for-all. Where do you draw the line, if any, on owning guns for protection? More people than burglars are shot with their own guns.

We are reminded of General George Patton's famous admonition, "Do something, dammit, even if it's wrong." He carried two pearl-handled revolvers himself. Whatever we do will make somebody mad. But if we don't do something, we might as well man the barricades and shoot it out at high noon. We're defending ourselves to death.

Whichever way it goes, we want 2-Gun Slinger on our side.

Go in peace, and love thy neighbor. Be careful not to shoot him.

The Dark Ages

1. Resolved that we build a new jail.
2. Resolved that the new jail be built out of the materials of the old jail.
3. Resolved that the old jail be used until the new jail is finished.

Canton City Council, Mid-1800

Our sheriff has announced that national accreditation is coming to the Sheriff's Department. This sounded like motherhood, but even this news has turned out to be controversial. We couldn't believe it! There is opposition.

So we travelled down to Pithecanthropus County in South Florida to check with Sheriff Percy Piltdown, who doesn't believe in accreditation. We thought we should get the other side of the story. And we did — both barrels.

Sheriff Piltdown was practicing quick-draws with his pearl-handled six-shooters as we entered his office. He was even quicker to state his case.

"I don't need no experts down here to tell me how to run my business. I know how already."

Before we could get in a question edgewise, Percy received a phone call from Deputy Chief Neanderthal Jones, who was interrogating a suspect in the basement. The Sheriff listened intently.

"Use the pliers," said the Sheriff over the phone and then explained to us that they were giving the suspect a manicure. "Good grooming is important. We like to take good care of our clients," he said with compassion.

The Sheriff began twiddling with his six-guns again, and three rounds went off through the ceiling accidentally. Then in came a call from Deputy Crow Magnon about another suspect in the basement. After a pause, the Sheriff advised, "Just tighten up the screws," explaining to us that they were doing some routine maintenance work on the plumbing. To emphasize the point, he fired six quick shots through the floor.

He apparently just missed somebody in the basement because another call came in. The Sheriff thought a few moments and then answered him again, "Just use the water one drip at a time." He then explained to us that it was his policy to provide showers for all his prisoners to keep a neat and tidy establishment. "We don't need nobody to tell us that," he said.

We tried to ask him about his policy on the use of firearms, but just then Deputy Chief Jones called again. Percy listened carefully and then answered, "If it's not helping, turn up the juice," explaining to us this time that they were blow-drying the prisoner's hair after his shower and shampoo. "We're very thorough. We just try to do everything we can to help our clients."

No sooner had he finished than still another call came in from Neanderthal Jones. At this point the Sheriff was clearly getting agitated.

"Now what?" he hollered. He listened for a while and then shouted, "Use the rack." Apparently the prisoner was not drying fast enough and had to be assisted mechanically, presumably by spin-drying.

As if to emphasize his point, he fired three more rounds through the floor. We're not sure if he hit anybody or not.

We left Sheriff Piltdown, fully satisfied that he needed something; but we didn't think it was accreditation. We were much relieved to know accreditation isn't necessary. Who needs a bunch of experts, anyway?!

Garbage In, Garbage Out

> Why do I rob banks? Because that's where the money is!
> *Willie Sutton, bank robber*

It could only happen in New York: somebody stole *eleven* city garbage trucks worth $110,000 each. It couldn't happen here. Or could it?

We went to visit our county Garbage and Sewer Commission to find out. We asked to see our favorite garbage and sewer commissioner.

"He's out riding shotgun on a garbage truck," they said.

"How come?"

"New York. It scared us to death. We take our jobs seriously around here. Garbage and sewage have got to go, come hell or high water."

The idea conjured up visions of cowboys and Indians and outlaws along with stage-coaches and covered wagons. We could just see our favorite garbage and sewer commissioner in all his glory and splendor dressed in buckskins and brandishing his trusty Winchester.

"Great Scott! This *is* serious," we said. "But what do you do at night?"

"Just like the good old days. We circle the garbage wagons and stand guard."

This all sounded like great security, so we went to the Pensacola Garbage Department just to make sure they got the message. The place was empty.

So we went to City Hall to sound the alarm. There was only one person on duty, a security guard.

"Where's the council? Where's the staff? Where is everybody?" we asked.

"Haven't you heard about New York? We're all out riding on the garbage trucks."

"Thank Gawd!" We were greatly relieved. "But what do they do for protection at night?" We suggested circling the wagons, just like the county.

"We have a much better system," said the guard. "Staff and council members are chained to the garbage trucks. They sleep with them all night."

We suddenly remembered our neighboring town. What about their garbage? We rushed across the bridge to find out.

"Have you heard about New York?" we asked.

"Piece of cake here," they said. "This isn't New York. We're a God-fearing, peaceful, law-abiding, non-alcoholic community. No sweat. We're in good shape. Stop worrying."

We agreed. What a relief! We had done our very best to get the message out. We felt fulfilled.

So far only one councilman, one commissioner, and two trucks are missing here. Hang on to your hats and your garbageman.

Excess Underwear

When the story of the holdup is spread, every citizen will seize his shotgun and rush to the street. Whistles will blow and all traffic, including pedestrians, will halt. Anyone besides the law officers who moves will be shot.

Reverend George Durham

Armed bandits have hijacked two truck-loads of Fruit-of-the-Loom underwear. This is a first in the annals of U.S. crime. Never before has anyone ripped off such a large batch of underwear.

Imagine the robbers' surprise and chagrin when they returned to their headquarters and opened up the truck to count their loot. The bandits thought they were hijacking cigarettes.

"What in hell will we do with two truck-loads of underwear?" moaned the number one robber.

"We can't wear that much underwear in fifty years," sighed number two. "The laundry bill would kill us."

"Well, we've got to do something with the damned stuff," declared number one. "Mount up, and we'll take it to our regular outlet and see if we can fence it."

They then offered the Fruit-of-the-Loom to their agent who regularly moved their stolen goods.

"Fruit-of-the-Loom cigarettes," exclaimed the fence, "I never heard of that brand before. Is this something new?"

"No, no, you don't understand. This is underwear," explained number one.

"Underwear! My God, what would I do with two truck-loads of underwear? I couldn't afford to be stuck with that much underwear at any price. Even the Russians don't need that much. You couldn't pay me to take the stuff. Take it someplace else; get it outta here!"

"What on earth can we do?" pleaded number one.

"Well, how about a gift to your favorite charity? You can take it off your income tax return."

"What income tax return? We don't file. We can't afford to, according to our accountant. We'd never get rid of the IRS. We've never reported any of our cigarette sales. It's all profit. The IRS would break us. Besides, they're not our favorite charity."

The fence threw up his hands. "Well, whatever you do, get that underwear out of here before I get caught with it. Fruit-of-the-Loom is probably looking for this stuff right now. I can't go to the pen for a dumb stunt like this."

"For God's sake, please don't tell any of our competitors about this. We'll be the laughing stock."

The robbers then inserted the following ad in the newspapers: "We are pleased to announce we now have in stock two fine truckloads of brand new major brand underwear, super-comfort, stretch bands, soft, absorbent, all-cotton, a variety of colors and sizes, both men's and ladies.' We will not be underpriced. We will pay $10,000 to the first party who will take this fine merchandise out of our inventory, the sooner the better. Don't ask any questions. Call us at our unlisted number."

Some crime doesn't pay! This one did.

Stickum & Gunk

> Get the thing straight once and for all. The policeman isn't there to
> create disorder. The policeman is there to preserve disorder.
> *Richard Daley, Mayor of Chicago*

Newsweek recently published an article on innovations in non-lethal instruments of battle to be used in place of real bullets. These non-bullets could be especially effective for riot control when you don't want to kill anybody. Presently police and army troops use rubber bullets on mobs, but rubber bullets can kill at close range. The latest inventions:

Beanbag Barrage. Police will shoot beanbags out of a contraption that looks something like a blunderbuss. The beanbags will knock you down but won't dent or kill you. This is certainly a much more humane treatment of rioters.

Gladbag Attack. Police fire a large sheet of wrapping material out of a cannon. This blanket wraps itself around a vehicle or group of rioters, envelops them completely like a sheet of plastic, and shuts them down completely. Police have to cut them loose with chain saws. Excellent results.

Stickum Attack. This gluey net, when fired out of a cannon, grabs the enemy like fly paper and renders him/her harmless. To release the captives, the police have to pry them loose with crowbars, chisels, or the jaws of life.

Gooey Gunk Attack. Police squirt this new plastic gunk on a rioter. The gunk immediately hardens and freezes. The rioter is paralyzed in time and space. He can't move a muscle until the police melt the gunk with a blow torch.

The thought occurred to us that these new devices would be excellent for use by politicians. For instance, if Speaker Gingrich needed to round up some Democratic votes on a bill, he could fire a stickum net into the House of Representatives on the minority side of the aisle. The ensnared representatives would then either vote right or stay stuck until they saw the light.

Locally we contacted a Democratic Senator to see if he could use any of these inventions in his next campaign. He was excited and delighted and said he certainly could.

"I can round up my voters with a stickum net or else hogtie my opponents with it. If I see a Republican getting ready to vote, my poll watcher can zap him with gunk until the polls close."

The *Newsweek* article also suggests that such new weapons might even be used in actual combat, but failed to say what to do in case the enemy is shooting at you with real bullets, and all you have is a load of beanbags. We can answer that one in a big hurry.

Duck!

BANG, BANG, BANG

> That lowdown scoundrel deserves to be kicked to death by a jackass—
> and I'm just the one to do it.
> *Congressional Candidate in Texas*

Just recently a burglar in South Florida broke into an elderly couple's home. The husband was disabled and sleeping separately from his wife. He woke up to find the burglar in his bedroom and hollered for help.

Mrs. Smith, his elderly spouse with failing eyesight, rushed to his aid with her pistol. She had never fired a gun before. She closed her eyes and pulled the trigger 3 times. Bulls-eye! One round went cleanly through the center of the burglar's forehead. He is cold, stone dead.

Guess what? The local prosecutor is now determining whether Mrs. Smith should be charged with murder. Whatever happened to the rights of the burglaree? Your house is supposed to be *your* castle, not his.

Former Attorney General Edwin Meese once brilliantly observed that "if a person is innocent of a crime, then he is not a suspect." But this guy was guilty, caught in the act, dead to rights. He is the suspect *corpus delicti* himself.

We are trying to imagine what it would be like to be politically correct these days during our crime wave. We think Mrs. Smith's ordeal should have gone something like this:

Burglar Jones breaks into victim Smith's cottage. Mrs. Smith hears the glass breaking and grabs her trusty .44 magnum with pearl handle.

She catches Burglar Jones in the act and levels her .44 at his large chest.

"Put your hands up," she orders.

"That's my line," he says.

"What are you doing in here? Confess before I shoot you," she says.

Burglar Jones falls to his knees, begging.

"Oh, Lordy, lady, please don't shoot. I was an abused child, and I'm being harassed by my ex-wife who won't pay me alimony. My family threw me out, I'm on drugs, and I can't get on welfare. I'm out on parole, my liberal congressman lost the election, and nobody loves me. I'm just trying to steal some money to buy drugs, and I'm trying to find some love. I'm a poor deprived citizen. The good life has passed me by."

"You poor lost soul," sympathizes Mrs. Smith. "What can I do to help you since you've been so abused?"

"Please, just put down your .44," pleads Burglar Jones.

"I might just as well," sighs Mrs. Smith. "If I shoot you, I'll go to jail."

Mrs. Smith lays down her .44. Burglar Jones picks it up and promptly shoots Mrs. Smith with it, dead.

The valet attendant from Jones' parole service picks up Jones and deposits him at the Country Club Manor Detention Center. Jones now enjoys hot running water, indoor plumbing, room service, gourmet meals, daily showers, wide-screen color TV, conjugal visits, tennis, golf, massage, and cocktails plus paid leave time away from his profession as a burglar.

Jones goes before the court and pleads insanity, self-defense, persecution, hunger, chronic addiction, harassment, husband abuse, unequal opportunity, etc. He says he was so high on drugs at the time that he doesn't even remember a Mrs. Smith, let alone shooting her. How sad!

In a fit of compassion the jury acquits Jones and awards him a pension, a medal, and a scholarship to Harvard University. Mrs. Smith's husband is required to make restitution for all the trouble his wife caused Jones.

Jones has now applied for workmen's compensation for stress related to his occupation. He should live so long.

Try the following actual live action story on for size. This is real political correctness:

Three teenage punks murdered a British tourist near Madison, Florida, and threatened his female companion. All 3 were caught. One was 13 years old.

The court sentenced him (now 15) to a curfew and 50 hours of community service, notwithstanding the fact he had 14 previous arrests and more recently attacked his counselor who was trying to straighten him out. How about a nice pat on the back for this punk? Shouldn't he be locked up in some jail, or under it?

Councilman John Bowman of Washington, D.C., recently commented on the high crime rate, "If crime went down 100%, it would still be 50 times higher than it should be!"

Is it any wonder?

The Sparks Flew

> How do I stand on capital punishment? I stand by the switch.
> *Governor Roy Roemer, Colorado*

Florida recently electrocuted Pedro Medina for murder. Fire flew out of Pedro when they threw the switch, so now some folks have complained that the state tried to sauté Pedro right then and there. They don't call the Florida electric chair "Old Sparky" for nothing!

Pedro Medina is the same nice gentleman who aerated an older lady with a butcher knife after she befriended him. Is anybody sorry about her? Not much as been said about that irrelevant question. There is much more concern about how we can execute criminals more comfortably so they will enjoy it more.

Now sparks are flying all over the capitol where the governor and the attorney general talk about getting the legislature to authorize lethal injection which is supposedly painless. One state legislator has already proposed the guillotine which he says is quick and painless. But how do we know? Sounds rather messy to us.

To get some idea of just what should happen we talked to Hugh Killem, the little man who actually throws the switch to fire up Old Sparky.

"What happened to Pedro?" we asked.

"Dunno. Guess he just blew a fuse. We musta overcooked him."

"Was it painless?"

"Dunno. Pedro never did say."

"What are the alternatives?"

"People are sending in lots of ideas. I'm an electrician myself, but I'm willing to listen."

"What are some options?"

"One guy wants to tickle 'em to death. We aren't sure that will work. Another nut wants to ply 'em with liquor and paregoric like the Heaven's Gate folks, then let 'em sleep in a garbage bag."

"Would that send them to a higher level?"

"Naw, I think these guys are going to a lower level."

"What else?"

"Another idiot wants to let them eat themselves to death on junk food. They would overeat at their last supper and die stuffed but contented," advised Killem.

"These are crazy ideas."

"You know it, but you ain't heard nuthin' yet. Some crazies have sent in some really screwy ideas. These won't work because we want the condemned to enjoy it just as much as their victims did."

"Such as?"

"One guy suggested we tie 'em down and make 'em listen to Al Gore speeches."

"That's torture," we gasped.

"I told ya. Worse yet, another one said we should put on earphones so they can listen to Tiny Tim."

"That's inhumane."

"Sure is."

"What do you recommend?"

"If it was me, I'd pin a bull's-eye on 'em and turn 'em loose between two teenage gangs. That's the end, brother."

"But you would be out of a job."

"Suits me. I can always get a job with the power company. I ain't interested in tickling some poor guy to death."

It sounds as though Old Sparky is still the best bet. They didn't kill their victims with kindness.

Outta Here

Our new management policy: One mistake and you're fired.

A fleeing criminal raced by outside, attempting to escape the police. A humble paint shop employee, opening up the shop and preparing to paint a car, heard the large commotion out on the street.

The painter, sensing his civic and patriotic duty, jumped into the customer's as yet unpainted car and took off in hot pursuit of the alleged crook. Finally catching up to the fleeing felon, he jumped out of the car and chased him on foot. He ultimately tackled the pursuee and sat on him until police arrived. A hero!

Our hero received loud plaudits from police, officials, and public alike. He was a man who saw his duty and did it. He received much acclaim and returned to work.

His boss demanded to know what was going on. When our heroic painter explained, the boss went ballistic. "You risked a customer's car without permission? You abandoned your post?" The painter tried to explain further, to no avail.

"You're fired!" shouted the boss. End of chase, according to the national press.

We were incredulous. How could such a thing happen to a conscientious citizen? We went out to conduct man-on-the-street interviews to feel the public outrage at such an injustice.

First we talked to a police chief and told him the story.

"Whadya think, chief?"

"I think it's criminal," he said.

"What's criminal?"

"Some crazy civilian interfering with the police. That painter stole a car and got in the police's way. We can't have every nutty hero running around doing police work. We might lose our jobs."

So much for the chief. Next we talked to a paint shop manager. What did he think?

"I'm 100% for the boss. If any of my folks pulled a crazy stunt like that, it's `Katie bar the door'. Out! Suppose that nut had gotten shot? I'd have to pay workman's compensation. Suppose he wrecked my customer's car? I get sued. My insurance premiums go sky-high. No way! He's outta here! Gone! Fired!"

Next we cornered a car owner who needed a paint job. What about our hero!

"No way. I'm supposed to have my car painted just so some goofy painter can go gallivanting around playing cops and robbers? Suppose the paint is still wet and runs off and drips all over the streets and the mayor? I'm ruined. Suppose he wrecks my car? There goes my insurance. Nuts! They oughta lock *him* up."

We couldn't really believe what we were hearing so we went to the jail and talked to a robber to get his perspective.

"I feel terrible. That painter had no right to mess around with a robber's civil rights. Robbers deserve equal opportunity to get away. You just don't interfere with somebody during his escape. They oughta put that painter in jail for keeps."

Citizens, beware. If you suddenly see your civic duty, don't!

So much for chivalry and justice. They're dead, dead, dead. Fired!

ON CULTURE

Morton Simon paid three million dollars for a painting by Raphael and
it's just got one coat of paint.
Unknown

It's not as bad as it sounds.
Mark Twain on Wagner

On Thanksgiving: Let all give humble, hearty, and sincere thanks now, except for the
turkeys. In the island of Fiji, they don't use turkeys; they use plumbers.
Mark Twain

Mr. Asquith was like a drunken man walking along a straight line—
the farther he went the sooner he fell.
Sir Edward Carson, Irish Politician

I saw the play under adverse conditions. The curtain was up.
George S. Kaufman

Is There Really?

If you don't ask, the answer is always No.
Owen Laughlin

One hundred years ago a young lady named Virginia wrote to a newspaper and
asked, "Is there really a Santa Claus?"

The editor fired back those immortal words heard round the world, "Yes, Virginia,
there is a Santa Claus." Things were more easily explained in those days before all the
new-fangled knowledge and all of our technology. Things are a lot more complicated
these days; so when another Virginia asked us about Santa Claus the other day, the conversation went like this:

Virginia: What's all this stuff about Santa Claus?
Us: Well, what about it?
Virginia: In the first place, how does one little fat guy in a red suit go all the way
around the world in an old sleigh with eight tiny reindeer in one night?
That's nuts.
Us: You're right. He can't make it so he traded in his sleigh on a new jet. That
really helps.
Virginia: Who are all those Santa Clauses I see running around town? Who's real and
who's fake?
Us: They're all fakes. Those guys are just Santa's helpers who hire out during
Christmas. They belong to the Santa Claus Union. They're required to wear
red suits and not to drink on duty. The real Santa rides in the Christmas
parade.

Virginia:	Then how come I watched two parades on TV, and they both had Santa Clause at the same time?
Us:	Oops. Your TV must have been out of order.
Virginia:	Out of order, my foot! That's some more of those fakes. I'm still confused. Does Santa do all the work at the North Pole? How can he make all those toys?
Us:	No way. Santa has Mrs. Claus and hundreds of elves who do the work. He also gets lots of stuff from *Toys 'R Us*.
Virginia:	How does he handle all the kids' letters?
Us:	Easy. He has a fax machine, and he gets E-mail.
Virginia:	How does he keep track of all those requests? There have to be millions of them. Nobody could remember all those kids, even on 3x5 cards. How does he file all that stuff?
Us:	No problem. He has computers. Nothing to it.
Virginia:	How did Santa get so fat?
Us:	Because you kids leave him all that milk and all those cookies.
Virginia:	OK. This year it's just Slimfast and Vitamin E. But I still can't believe Santa covers the whole globe by himself even in a jet.
Us:	He doesn't. He uses Federal Express.
Virginia:	Santa is too fat to fit in our fireplace. Does Federal Express come down the chimney?
Us:	No, they just use the front door.
Virginia:	Why isn't Santa female? Aren't we supposed to have equal rights?
Us:	Virginia, don't ask. Just believe. We do.

Union Santas

> Santa Claus used to work for Federal Express
> *Great Quotations, Inc.*

Santa Claus had a bad year.

First, two Santas showed up in the same Christmas parade. The kids went into shock. How can you explain this snafu to a child? Santa called in counselors to cope with the trauma. Next, kids will wonder about the stork.

Then a hunter shot at Santa's reindeer by mistake as they came in for a landing. He nicked Donner and Blitzen.

The Air Force couldn't identify the sleigh flying on instruments and fired missiles. Santa almost lost Dancer and Prancer to flak.

Then came the ultimate insult of all: a preacher called for a boycott of Santa Claus. A *boycott*! A boycott could cause a recession. This very thought drives Alan Greenspan nuts.

Now the Amalgamated Santa Claus Helpers Union threatens to strike.

We can hardly believe all of this bad news. We went to visit a member of the union. We found one sad-looking Santa at a bar on his lunch break.

"You look lonely," we sympathized.

"You're dang right. I don't dare get caught with another Santa. It drives the kids nuts. They think I jump from one store to the next.

"Why strike?" we asked.

"Because the Santa Claus management don't give a hoot about us. We have to sit around with a bunch of kids with runny noses. Yesterday one wet all over my lap."

"But Santa has to have helpers," we reminded him.

"Let him take his own orders. We do all his work, and he gets all the glory. He rides around in that sleigh trying to look like Queen Elizabeth."

"That's not fair," we said.

"You don't know the half of it. He doesn't pay us enough to live on while he sits around on his big fat bottom at the North Pole with a big fat salary. His elves do all his work. I'm unemployed most of the year."

"But you can draw your unemployment," we offered.

"Humph! I can't live on that! Look at me. I'm starving to death. I'm so skinny the government will pull my license. You have to be fat to play Santa. I can't find enough padding so I'm in here drinking a beer to try to fatten up." He ordered another beer.

The union attitude was ominous. We went to warn Santa. "Did you know your helpers are threatening to strike?"

"That's fine with me," said Santa. "All they do is sit around and drink beer and take orders from all those greedy kids. Those fake Santas will promise kids anything just to get them off their laps in a hurry. They load me up with all these orders. I can't keep up."

"What will the kids do?"

"They can fax me. I'll put a limit on what they can ask for. Most of 'em are bad, anyway. One kid wanted a machine gun and 2,000 pounds of ammonium nitrate."

"What was that for?"

"The kid threatened to blow me up if I didn't deliver. Kids have gone nuts!"

"There'll be a lot of unhappy kids out there," we warned.

"That's too bad, but I take all the risks. I've already been shot at, sued, and boycotted. I may just shut down."

"This is tragic," we moaned.

"I'm sorry, but the union guys are too big for their Santa Claus suits. Tell 'em to go suck on a candy cane for all I care."

Needless to say, we were demoralized over this sad turn of events. There may be no Xmas next year.

Happy New Year anyway.

Xerox Santas

> Murphy's Law: If something can go wrong, it will go wrong.
> Sullivan's Law: Murphy was an optimist.
> *Great Quotations, Inc.*

A possible Christmas tragedy has been narrowly averted this year: in the nick of time the town council in Jay, Florida, passed an emergency ordinance prohibiting any more than one Santa Claus from appearing in this year's Christmas parade.

It seems that last year four Santas turned up in the Jay parade, much to the consternation of Santa's little constituents. The children love Santa, but just one at a time.

Santas are forbidden by union rules to congregate. It is a cardinal sin in the trade for one Santa to be caught together with another Santa in plain sight of their clients. They are supposed to work in shifts.

We clearly remember one harried mother who tried to explain 13 Santas in one four block area of downtown Pensacola. Then suddenly ahead of her two Santas on coffee break appeared at the corner of Main and Palafox Streets at the very same instant because of a mix-up in scheduling at union headquarters.

Horrified, the mother grabbed up her small boy, lest he glimpse such a compromising faux pas, and darted into Trader Jon's for safety. There, propped up against the bar on coffee break, were three more Santas, each with a Bud Lite.

In that one awful instant the sacred Christmas mystique of Santa was unveiled for one small American. An American tragedy!

Under such circumstances how could anyone possibly explain to Virginia that "Yes, there is a Santa Claus." Just one?

As the mother rushed her son from the scene of disaster, we heard the little boy ask her, "Now what about the stork?"

Charge

> A nickel ain't worth a dime any more.
> *Yogi Berra*

The Christmas season is a time of peace on earth. Or is it?

Violence has already raised its ugly head during this period of joy. Consider the day-after-Thanksgiving sales. People lined up before dawn to get into the stores. Opening time was pure bedlam.

We talked to Mark Downs, manager of a local retail store. Mark was bloody and bruised from head to foot.

"What on earth happened to you?" we asked.

"It's those damned pre-Xmas sales. I was up all night getting ready to open. People lined up all night to get in early."

"What time did you open?"

"We didn't open. They broke down the doors at 5:30. We couldn't hold 'em off. It was worse than a cavalry charge."

"Where were you during the stampede?"

"I was just inside the front door when the mob broke in and attacked us. It was a regular tidal wave. I was shoved all the day way into intimate apparel at the other end of the store. I ended up behind the girdles. It was awful. Women fighting over underwear. I was stuck in foundations all day. I couldn't fight my way out."

"It sounds awful."

"It was terrible. People got trampled. We called the police, but it was too late. Look at me. I'm seriously wounded."

"What on earth were people thinking of?"

"They were fighting for those damned Furbies. They went nuts over a bunch of talking dolls. I hate 'em. They talk all the time. You can't shut them up."

We were disheartened at this sad state of Xmas brotherly love. Apparently all retail outlets suffered similar riots. We wondered if financial institutions fared any better. Apparently not.

At a bank in Stamford, Connecticut, Edward Aragi, an irate depositor, dragged a dead deer into the lobby and plopped it down on a customer service person's desk where a sign read "The buck doesn't stop here." Unfortunately that's exactly where the buck landed. We called customer service person Robin Banks to find out what happened.

"We screwed up," admitted Robin. "We bounced the guy's check he had written on his Xmas savings account. Whew! Was he mad. He flipped completely and threw this big dead deer at me. Blood and guts and fur all over the place and me too."

"Who was right?"

"He was right about his check, but you can't just throw dead deer into a bank. What will all the other customers think? He could get six months for this, you know."

"Where's the deer now?"

"The head is mounted right over the cashiers," she said proudly. "and we're giving away venison for Xmas. We had to try something to get our customers back.

Santa Claus won't believe all the trouble he's caused already. Maybe the new year will bring me good will toward men and women.

You're Coming in Garbled

Tenses, Gender, and Number: for the purpose of the rules and regulations continued in this chapter, the present tense includes the past and future tenses, and the future, the present; the masculine gender includes the feminine, and the feminine the masculine, and the singular includes the plural and the plural the singular.
California Code

We wish to show how important good, correct, and simple English actually is. Straight thinking is also necessary.

Misplaced modifiers, mangled syntax, and foggy thinking can lead to unexpected results. These results can be demonstrated by the descriptions in 30 words or less that accident victims actually submitted to their insurance companies.

Try the following accident reports for fun and garbled thinking:

- Coming home, I drove into the wrong house and collided with a tree I don't have.
- The other car collided with mine without giving warning of its intentions.
- The guy was all over the road. I had to swerve a number of times before I hit him.
- The gentleman behind me struck me on the backside. He then went to rest in the bushes with just his rear end showing.
- In my attempt to kill a fly, I drove into a telephone pole.
- The accident occurred when I was attempting to bring my car out of a skid by steering it into the other vehicle.
- I was on my way to the doctor's with rear end trouble when my universal joint gave way causing me to have the accident.
- To avoid hitting the bumper of the car in front, I struck the pedestrian.
- My car was legally parked as it backed into the other vehicle.
- An invisible car came out of nowhere, struck my vehicle, and vanished.
- I told the police that I was not injured, but on removing my hat I found I had a fractured skull.
- When I saw I could not avoid a collision, I stepped on the gas and then crashed into the other car.
- The pedestrian had no idea which direction to go so I ran over him.
- I was thrown from my car as it left the road. I was later found in a ditch by some stray cows.
- The accident happened when the right front door of a car came around the corner without giving a signal.

- The telephone pole was approaching fast. I was attempting to swerve out of its path when it struck my front end.
- Walking down the hall the slippery floor came up suddenly and cartwheeled end-over-end.

So be careful. Think straight. Write straight. As General Von Moltke put it, "If an order can be misunderstood, it will be misunderstood."

Gobbledegook

> Every nation must have its own traditional language as a primary
> language which if it was not English is not likely to be.
> *George C. McGee*

The English language can be simple and effective until it gets into the hands of the wrong people. Then it becomes "unplain" English.

Confusion abounds in government and legal forms and various technical instructions as well as pompous declarations by bureaucrats and elite intellectuals. The Internal Revenue Service does not write simple English, a practice followed closely by lawyers, insurance policies, tax laws, doctors' prescriptions, and instructions for assembling children's toys on Christmas Eve.

Even simple instructions and questions on employment forms can be confusing and misunderstood. One poor lady, filling out an application for a job, came to the question —SEX?. She thought for a long time and then finally wrote, "Yes, once in Atlanta." Technically she was absolutely correct.

Far more serious is the gobbledegook invented by bureaucrats to sound important. They say "at this point in time" when they mean "now" or "at that point in time" when they mean "then." They "outsource," and they either "micromanage" or "macromanage," whatever that means. Everything has to be "viable." "Infrastructure," whatever it is, is everywhere. And bureaucrats do everything with "specificity." Bureaucratese is the worst of sins and originates in Washington, D.C.

The crowning achievement in Washington, D.C., came in the form of an analysis of a government questionnaire asking about drivers' attitudes toward trucks, even though we already knew the answers. Here is what the government analyst concluded:

"This particular folly was justified on the grounds that previous research had focused on aerodynamic and human factors. The thrust had been on the physics of interaction and the psychophysical performance of drivers. Obviously, the cognitive or mentally evaluative dimensions of such vehicular interactions or potential interactions was (sic) lost in error variance or not considered at all in formulations of aerodynamic functions and driver performance factors."

Try translating that from the Greek. That little analysis cost the taxpayers $220,000.

George Will points out that in 1966 English teachers decided not to stress punctuation, coherence, or grammar — just write *something* down *somehow*. "This new era of enlightenment was not a matter of putting things into students, but of letting things out." Good Lord!

In her "Essay on why Johnny can't write," Heather MacDonald points out that "students... have been told in their writing classes to let their deepest selves loose on the page and not worry about syntax, logic, or form..." No wonder students are learning "unplain" English. They take it with them into real life and government.

In attempting to justify the free-for-all philosophy and the decline in plain English skills, a pompous professor expounded as follows: ..."post-process, post-cognitive theory... represents literacy as an ideological arena and composing as a cultural activity by which writers position and reposition themselves in relation to their own and others' subjectivities, discourses, practices and institutions."

Figure that out! Only a professor could understand such gibberish. He would do well in government.

"Unplain" English is catching and can eventually catch up with you if you're not careful. An ambitious young employee, trying to impress his boss, proudly reported that he had "matriculated on campus and practiced nepotism with his sister." His boss fired him for immorality.

Keep English plain and simple!

We Never Had It So Good?

> One of the hardest decisions in life is when to start middle age.
> *Great Quotations, Inc.*

We just had another birthday which we didn't need. Birthdays are alright for kids, but at 72 who needs to celebrate one more year lost? Unfortunately you can't hold back the sunset.

Maybe birthdays are meant as a good time to look back and remember the good old days. Just for fun, compare the cost of living in 1923 with 1995:

	1923	1995
Gallon of milk	$.54	$2.29
Loaf of bread	$.09	$1.59
New car	$500.00	$12,371.00
Gallon of gas	$.11	$1.19
New home	$7,400.00	$106,100.00
Income average	$1,265.00	$19,587.00

Maybe we're better off today, but a loaf of bread is still a loaf of bread at 17 times the 1923 price. Is inflation worth it?

In the 1920's teachers worried about cutting in line, talking in class, and chewing gum. Today teachers worry about drugs, violence, theft, and weapons.

In the old days we went to neighborhood schools. No buses. We walked or rode our bikes to and from grade school, including home for lunch.

We took streetcars or city buses or drove to high school carrying peanut butter and jelly sandwiches. Parents were responsible for getting us to school one way or another and feeding us while there.

Divorce was rare, not a fact of life. You got married BP (before pregnancy). Today people live together without benefit of matrimony, and the divorce rate approaches 50%.

Drugs were something you bought at the drug store. Now, mind-curdling drugs are sold in kindergarten and on most street corners.

There were no wonder drugs in the old days, mostly just aspirin and castor oil. Doctors were much scarcer, but they made house calls. Medicine and doctors were less expensive, but we did not have all of the current miracle medicines, machines, and shots. We all got measles, mumps, chicken pox, and whooping cough. And there was the dreaded polio.

We tried prohibition of alcohol (the non-medicinal kind). Bootleggers replaced liquor dealers. Al Capone got rich along with lots of other gangsters. People drank whatever they could get their hands on, including bathtub gin. Homemade brew blew up in basements. Prohibition didn't work.

Cars didn't cost much but looked funny: square, hand-cranked models, no streamlining. They didn't go very fast; roads weren't so good either. No interstates, just 2 lanes and lots of ruts and flat tires.

Airplanes flew low and slow, no jets, just propellers; air mail flew in windy, open cockpits with pilots in goggles. No radar, instrumentation, automatic landing, autopilot, direction-finding, computers, etc. Pilots flew along roads or power lines or by the seat of their pants.

Lawyers were scarcer, and everybody was not suing everybody else daily. Today the U.S. has 800,000 lawyers, over half of the lawyers in the entire world! We have twice as many per capita as Britain, four times as many as Germany, twenty times as many as Japan. We have a lawyer in every ambulance.

Government still did dumb things in those days but was smaller and less dangerous and couldn't do nearly as much damage.

1929 was a bad year. Everybody who went broke jumped out of a window. There were no mutual funds, and no pesky brokers calling during dinner. There was much less inflation, lower interest rates, no national debt, no derivatives (whatever they are).

So there was good news and bad news in those days too. Maybe some things about the good old days weren't so great after all, but birthdays are still a drag unless you consider the alternatives.

Here's to a few more!

Ears & Spears, Etc.

It's more than magnificent. It's mediocre.
Samuel Goldwyn

These are strange times, and they seem to be getting stranger with every passing hour.

The most current strangeness is *piercing*. This fad concerns the piercing of various body parts with rings of various sizes and shapes and descriptions.

We're not just talking earrings here. People, mostly young, are piercing every conceivable part of their anatomies they can think of: ears, lips, noses, nipples, belly buttons — you name it. Earrings have always been a standard accouterment for women since the cave dwellers, but not a dozen earrings in the same ear.

We have tried to unearth some historical precedent for this strange practice. We recall one African tribe, the Ubangi, whose women pierce their lips and then stretch them around flat wooden disks so that their lips resemble large waffle irons.

In fact, these ladies can consume a whole waffle in one gulp, which seems much more practical than simply piercing with adornments for ornamental purposes. But how you kiss a waffle iron we simply do not know.

Going farther back in time, we recall the cave people. They were constantly piercing each other with spears but scarcely for the purpose of beautification. These piercings were generally terminal. As somebody said, "You only die once but for such a long time."

There is no record of piercing among the apes or the dinosaurs. Apparently they had more sense than we give them credit for.

We wonder how most modern parents react to such self-mutilation. In our day the only piercing would have been the screams of our parents when they first beheld such an atrocity. Our rear ends would then have been warmed, but not pierced.

The most amazing piercing today takes place in the National Football League. We see massive steroidal monsters with bulging, rippling muscles stalking the pro gridirons. We also behold these same hired killers wearing dainty earrings, long hair, and pony-tails. What's next — hair ribbons? What on earth? The whole scene is incongruous with paid mayhem. These behemoths are anything but dainty.

We are not sure whether modern society is reverting to cave people or not. This is reverse evolution. We hope this piercing craze does not return us to the days of spears and inter-tribal warfare. God forbid!

Suppose that current political mud-slinging turned to spearing. This could be serious, but it also might be beneficial if the candidates would just spear each other and put an end to our agony of political ads.

Privy Progress

> Too much of a good thing is wonderful.
> *Mae West*

The news is almost always bad, but then every once in a while a fun item shows up amidst the gloom and doom just in time to make your day.

We heard recently about the Privy Preservation Society in Philadelphia. No joke! These folks are deadly serious. They uncovered a solid brick colonial outhouse and are restoring it for posterity. Historic Pensacola, take note: we may have some of these hidden treasures here.

You wonder why somebody who went to all the trouble to build a solid brick privy didn't attach it to the house for aesthetic purposes. The sheer beauty of the thing! An attached outhouse can't be tipped over, but of course neither can a solid brick model. The non-tip feature takes all the fun and games out of outhouses.

The greatest joy that could come to the neighborhood clowns from an outhouse was tipping it over. That was the prize prank in the good old days, particularly if the privy was occupied at the time. The occupant found him/herself in an untenable position. It was no laughing matter except for the tipper — the tippee was never amused.

Besides brick, which was certainly the rare exception, there were many different models of these very vital facilities. One common feature: they always had a half-moon on the front door for identification purposes. That was standard. There was no mistaking the outhouse for something else.

The most common model was the two-holer on a standard two-hole chassis. This model was wood-frame and came with a variety of wooden siding or shingles. A rack to hold the Sears catalog was also standard. The invention of the outhouse actually led to the publication of the Sears catalog; the two were inseparable. A rack for a shotgun was optional. This was for defensive purposes in case of attack.

There was also a luxury model, a two-holer on a four-hole chassis, spacious, lighted, a real dream. Such a privy might be stuccoed outside with wallpaper or fur-lining inside, insulated, with a cooling fan and heater for extra comfort. There was also an emergency exit. These luxury units compared favorably with today's finest condominiums.

Then there was a convertible model. You could let the top down just like a Cadillac on bright sunny days for added enjoyment. Because of its low center of gravity, this model could not be easily tipped over. Also the visibility was better; attackers couldn't sneak up on it as easily.

The most ingenious hoax we ever heard of was also the most scientifically advanced. It was also downright diabolic. The pranksters installed a loudspeaker underneath the privy and located a concealed microphone at a safe distance from the scene of the crime. Just as the occupant had settled down peacefully with his catalog, a booming voice blared out from below, "Mister, would you mind moving over one seat? I'm painting down here." The results were so traumatic we would not be able to describe them here. The culprits were never apprehended.

All the fun ended when privies moved indoors...but not quite. History records that the first lady ever to see an indoor commode thought it was a newfangled washing machine. She put in four shirts and pulled the chain. The shirts were never seen again.

Time marches on!

Beethoven in the Nude

> Clothes make the man. Naked people have little or no influence in society.
> *Mark Twain*

Police report the arrest of a man who was playing his accordion in a laundromat. The charge? Disorderly conduct.

How, you ask, does playing music while doing your laundry amount to misconduct? Well, it seems that said musician was absolutely raw, stark, buck naked except for his accordion. We wondered what on earth would possess a gentleman to perform such an act.

We visited nude accordionist Francois Des Corde, a Frenchman, in his cell at the jail. By this time he was covered with a blanket and was quietly fingering Beethoven on his instrument.

"What on earth happened?" we asked.

"It's a long, sad story."

"Try us. Why were you undressed?"

"I had to wash my clothes, and the only ones I had were the ones I had on. My wife got all the rest. I wanted to get into the washing machine with my clothes on, and then I could drip-dry afterward; but there wasn't enough room in the machine. Besides, insurance regulations prohibit bathing inside a washing machine. So I took off my clothes and put them in, but I wasn't exposed. I still had my accordion on; there were no ladies present."

"Why didn't you let your wife wash them at home?"

"What wife? She hates accordions so she took up with a rock band. I can't play rock or rap or heavy metal."

"Why didn't you go home to wash?"

"What home? My wife got the house along with the rock band. I can't stand to go back and listen to that awful stuff. And they don't want an accordion."

"Where are the rest of your clothes?"

"My wife gave them to the rock band. Then some gang ripped off everything else I had except my accordion."

"Can't you get on with Lawrence Welk?"

"I tried, but he's already full of accordions."

"This all sounds pretty innocent to us so far. What happened next?"

"Well, some ladies came in to do their washing, and I played Beethoven for them. They seemed to enjoy the music. I still had my accordion on, and they didn't know I didn't have any pants on."

"What went wrong?"

"I forgot and stood up and took my accordion off to go get my clothes out of the machine. That did it! All hell broke loose. The ladies screamed, and somebody shouted 'Packwood!' They called the police."

"What does this have to do with Senator Packwood?"

"Well, you know, he's had some trouble with ladies in the Senate. They're pretty paranoid about creeps and sex maniacs. But I didn't kiss anybody or pinch anybody."

"You're not Senator Packwood."

"I wish I were. He got to resign and go home and write a book. Here I am in jail with no clothes, just a blanket and an accordion. I don't even keep a diary. Back in Paris everything would have been O.K. The French understand these things."

"Did you explain all this to the police?"

"Well, I put my accordion back on, but they didn't buy it. I had to take my accordion off while they handcuffed me, and the ladies started screaming again. So here I am. Woe is me."

"Where are your clothes now?"

"Still in the washing machine. They're impounded as evidence. I'm going to court in this blanket, and I'd better wear my accordion too. What will the judge think?"

"What about the ladies?"

"They're testifying against me, but they did seem to like my music."

The judge listened to Francois' story but didn't buy it. The ladies did recommend mercy. The judge sentenced him to 30 days in the laundromat playing his accordion, fully dressed, then 2 years probation with Lawrence Welk, also fully dressed.

Francois is on the road to recovery except for his house, his wife, the rock band, and the rest of his clothes.

Older is Better

We had our last car for 19 years.
We grew so attached to it, we couldn't trade it in. We had it put to sleep.
Bob Orben

Our car is only 15 years old.

We drive a 1983 Oldsmobile, and it's a beauty. It has only 120,000 miles. It's big and long and heavy and gets 5 gallons per mile in town, even better on the highway.

But our granddaughter is disgusted with us; she can't understand why we don't buy a new car. The answer is simple: they're too darned expensive for what you get. They're too small, and they all look alike, squashed down low in front and sticking up in the back with a big rear end. They all look as if they're sliding downhill, and you can't tell one brand from the other.

Our car is big and square and stands up tall and proud. It looks like a box. No mashed down front end for streamlining; we don't need a race car. It doesn't have any computers; almost everything works by hand; the brakes don't squeak too badly, and the parking brake almost works. It's broke so why fix it?

The upholstery on the ceiling is dimpled where it is detaching from the roof, but who cares? You don't drive looking at the ceiling.

We inherited the tradition of driving old cars from our father. He drove a car, and then he drove it and drove it and drove it and drove it some more. He was an attorney, and he was penurious (tight). We also inherited that tradition.

One morning dad started out with his briefcase to try a big court case. He was nervous and in a hurry. Suddenly we heard a loud splintering noise in the back of the house. Dad backed out of the garage without opening the garage doors. There was kindling wood scattered all over the alley, but he never looked back.

On the way downtown there is a long, steep hill. Halfway down his brakes failed. At the foot of the hill is an infamous, wild intersection known as Seven Corners. Seven streets intersect there.

He sailed through Seven Corners at 70 miles per hour, no brakes, wildly waving his arms, horn blaring, traffic scattering, pedestrians fleeing for their lives. He was able to coast all the way to the Buick dealer. He stepped out proudly and announced, "I think I need a new car." He did, but we aren't ready to trade ours yet.

We inherited our dislike of computerized cars from our father-in-law, Leon Clancy. He hated all electronic and computerized gadgets; he referred to them as "hemorrhoid switches," and worse. To Leon, "automatic" meant you can't repair it yourself.

He traded with Mitchell Motors for his new cars. It was a classic matchup when Leon and Howard Mitchell sat down to negotiate for a new car. Clancy was a tough mule trader with sawmill experience. When the deal was finished, Howard was crying and Clancy was smiling.

One day after church Leon's new car wouldn't start. "Damned hemorrhoid switches," he screamed so everybody in church could hear. Leon pitched a fit and called Howard in language unfit for the sabbath or any other time. The congregation and preacher escaped, holding their ears.

"Come out here and get this *#@!?!*#! contraption," Leon shouted over the phone. Howard arrived in a hurry.

We took our own prize car to Super Sam the Used Car Man to get an appraisal. "Get that piece of junk outta here," shouted Sam.

So we took it next door to the junk yard. The junk man weighed it and gave us a price per pound. But we're not about ready to sell. We aren't going to trade this beauty for one of those dinky little modern cars.

The trouble with people is that they don't appreciate a real antique when they see one.

Smoke, Smoke, Smoke

> It's easy to quit smoking. I've done it hundreds of times.
> *Mark Twain*

Bill Clinton and Bob Dole are fighting over cigarettes. Smoking is now a political issue.

Clinton says he never inhaled, and Dole says milk, whiskey, and french fries may also be addictive so those two characters are certainly not qualified to settle the question. The question is addiction.

We know a lady named Paula Puffin who is a heavy smoker. We went to see her to get an expert opinion.

"Paula, how long have you been smoking?"

"Eighty years."

"My goodness, how old are you?"

"Eighty years."

"Good gracious, you started pretty early."

"Yes, I was weaned on cigars."

"Are you addicted?"

"Absolutely not. I'm in the process of quitting again right now."

"How do you do that?"

"It's easy. You buy nicotine patches and plaster one on your back every day. Nothing to it. The urge is gone."

Paula excused herself and went into the bathroom. Pretty soon smoke began to seep out from under the door. We were concerned.

"Are you all right in there?"

"Sure, I'm just great."

"We smell smoke."

"That's not smoke, just steam from the sauna."

It smelled like smoke to us. The smoke got thicker.

"Something's burning in there," we hollered. "Should we call the fire department?"

"No problem. Just get the hose."

We opened the door. The smoke was so thick we couldn't find Paula. We squirted the hose anyway and put out the cigarettes.

"We're afraid you're backsliding," we pointed out when we found her.

"Just a temporary relapse. I'll be fine with a new patch and a wad of nicotine gum. I'm not addicted to cigarettes."

We weren't sure about that. The next day the newspaper reported a four-alarm fire at Paula's place. Paula escaped; she was barely singed. We called to be sure she was all right.

"Sure, I'm fine. I'm not addicted. I don't need to smoke. If I feel the urge coming on, I just go to a bar and sit in the smoking section. There's always plenty of second-hand smoke to go around for everybody."

We called Paula's doctor to confirm her condition. He assured us she was fine. She is no longer addicted to cigarettes, he said.

"How are her lungs?" we wondered.

"Fine," said the doctor. "Just full of soot."

As Dizzy Dean once said, "Son, if ya really done it, it ain't braggin!"

Paula really done it. She has kicked the cigarette habit. Nothing to it. Now she is addicted to the patch.

Peace Pipes & Scalps

> We're completely surrounded. Don't take any prisoners.
> *General George Armstrong Custer*
> *at the Battle of the Little Big Horn*

We love the Indians, our Native Americans, that is. They're patriotic and tough and independent.

You will recall that the state of Florida tried to stop the Seminole Indians from bungee-jumping on their reservation. The Seminoles said "nuts" and kept right on jumping.

The governor sent his attorney general and other bureaucrats to enforce his edict. They pleaded and coaxed and threatened and tried bartering with trinkets.

"We no want them damn beads. We want wampum," said Chief Wigwam. The Seminoles threatened to burn the attorney general at the stake and scalp the rest of his crew so the Indians could sell their scalps to Yankee tourists.

Now comes the sad story of the Ojibwa tribe in Michigan. The tribe has split up into two opposing war parties and is fighting a civil war over who should get the profits from the bingo games held on the reservation. Bingo on Indian reservations is big business; we're talking big money here.

At first the warfare was mild, just some insults and sign-waving. Then, after studying up on our Civil War, matters heated up considerably. The war started with both sides throwing bales of bingo cards at each other.

Then they started throwing parts of the bingo machine back and forth, then finally the whole machine. Parts and pieces flew all over the reservation. Now it's bows and arrows, Remington rifles, and lawyers. This is a serious fight.

An American peace-keeping mission went to Michigan to try and make peace among the Ojibwa. The peace-keepers first met with Chief Ognib (that's bingo spelled backwards), the leader of the rebel force which is trying to get its hands on the bingo profits held by the governing party.

"We're being swindled," declared Chief Ognib. "They're big, wasteful spenders. We want a balanced budget, term limits, charter schools, more character, and less tax."

"You sound like Republicans," said the peace missionaries.

"Absolutely. We're Dole people. The Democrats threw us out. They lied. They need scalping."

"How about smoking a peace pipe?" the peace-keepers asked.

"They can take their peace pipe and stuff it," growled Ognib. He shot several burning arrows at the bingo hall.

The mission gave up on Ognib and went to see Chief Shooting Bull, the leader of the opposition. He was calling a bingo game. The peace mission related Ognib's complaints.

"Under the O, 26; under the B, 64," announced Shooting Bull. Then he glared at the mission. "Ognib's desperate. They're just grasping at scalps. We'll fix them the same way we fixed Custer. We're the party of compassion. We care about people, especially the ones who play bingo."

"Can't you share the wealth?" asked the missionaries.

"Nuts to that. We're in, they're out. Ognib's slinging mud and arrows at us. He needs scalping."

"Can't you smoke a peace pipe?"

"He knows what he can do with his peace pipe." Shooting Bull sharpened his tomahawk and went off to scalp Ognib.

The peace mission has failed. It looks as though the Ojibwa will fight to the bitter end. It seems as though we other Americans have poisoned Indian culture with our white man's politics. No wonder Sitting Bull mowed down Custer and his troops.

War Is Hell

Here lies Captain Ernest Bloomfield. Accidentally shot
by his orderly, March 2nd, 1879. Well done, good and faithful servant.
Epitaph on grave of British Soldier

Down through the ages man has fought great historic battles: Thermopylae, Waterloo, Bunker Hill, Gettysburg, Little Big Horn, Midway, Normandy, and Desert Storm.

Now we add another great battle to the history books. On or about 19 May 1997 there occurred in Tampa, Florida, what shall now and henceforth be known as the McDonald's Massacre.

It seems that a joyful group of youngsters were celebrating a happy birthday for one of their little friends in the party room at a McDonald's restaurant. Doting parents attended to supervise the festivities. We think this is what happened.

Before the party ended a second group of parents and children arrived to celebrate a second happy birthday.

The second party waited on the first party to finish ... and waited ... and waited. Impatience set in. Then frustration. Then anger. Finally, words were exchanged.

"It's time for you people to get the hell out of here. You're late. It's our turn," announced the senior parent of the second party.

"Oh, yeah, who says?" asked the senior parent of the first party. "We paid for this party, and we'll leave when we're damned good and ready."

"Oh, yeah, I says who!" answered party number two. "You're outta here, *NOW!*" With that he picked up a handful of french fries and hurled them at party number one.

"That does it," bellowed number one. Back came half a chocolate birthday cake launched directly at father number two. It missed the main adversary but caught mother number two in the middle of her forehead. She was aghast, then mad, then furious. She fired a chocolate milkshake directly at enemy number one. Milkshake showered the entire party.

An irate number one mother, drenched in chocolate milk and draped with chocolate cake, triggered a barrage of M & M's which peppered the aggressors with shrapnel. Now the children on both sides entered the fray to reinforce their parents.

Fries, candy, chicken, hot dogs, hamburgers, soda flew in all directions. Needless to say the manager became slightly agitated at the pitched battle being waged in the middle of his restaurant. He called for a cease fire, whereupon he was pelted with a storm of groceries.

The battle grew fiercer. We are reminded of Custer's immortal last words: "We're completely surrounded by the enemy. Don't let a single one escape."

With most of his menu now covering him from head to foot, the manager retreated to the parking lot and sent out a distress call to 911. A squad of police arrived in an attempt to quell the riot. They were soon covered with garbage. Mediation failed.

The warring parties covered the police with the rest of the menu which now was knee deep on the floor. The fury of the parents on both sides knew no bounds. In a rage they physically charged the police. The police handcuffed five so-called grownups and led them away, dripping junk food and charged with assault and battery on police officers.

Out of ammunition, both sides retired behind their respective lines. Party number one retreated from the battlefield. Party number two, licking its wounds, regrouped, celebrated, sang happy birthday, and went home still mad.

McDonald's manager is in a psychiatric ward. War is *still* hell.

Smoke, Chew, Sniff

I don't want want to tell you any half truths unless they're completely accurate.
Dennis Rappaport

Cigarettes are out!

The Surgeon General has raised so much hell about cigarette smoke and cancer that nicotine addicts are looking for alternatives. What could those be?

To get some answers we visited the Pink Powderpuff Pool and Poker Parlor. Inside was a cigar bar and lounge. La Belle de La Flambé, hostess and bouncer, met us at the door.

"You can't come in here," she warned. "This is for women only."

"But we're the press, and we need a story. We would like to interview some of the ladies," we requested.

"We're not ladies; we're cigar smokers. And I ain't no lady, period." We could believe that statement.

She finally did let us look around. We tried to find some customers in the dense cloud of cigar smoke. We could barely make out dim figures coughing and gagging in the milky fog. We watched one delicate little old lady chomp down hard on a big, long, fat, round, black cigar. She bit it in two and swallowed half.

Another lady was puffing so hard her whole cigar was on fire. She was unable to talk. Suddenly somebody hollered "Fire," and everybody put on a gas mask. They pulled out the fire hoses. We escaped before the place burned down completely.

Women smoking cigars? What are we coming to? Cigars belonged to a man's world before equal opportunity and all those other laws and regulations brought in the women.

We are familiar with male cigar smokers. Out father-in-law smoked big, long, fat Webster Fancy Tail cigars. You could see the red tip of his long cigar long before he even turned the corner, and you can whiff a cigar long before that.

So what about the men? What are male cigarette smokers doing to break the chain? We went to see Smokey Smith at the Downtown Sports and Cigar Bar for men only.

"What's up, man?" we asked.

"Cigars are out for us. The women bought them all up. We're chewing tobacco," said Smokey.

Then we noticed that the bar was ringed with spittoons. Smokey took a hefty chaw of Copenhagen and spat; he scored a direct hit 15 feet away. As the other customers chimed in, they played a tune on the assorted cuspidors. Ping! Pong! Ding! Dong! Perfect bell ringers. The Anvil Chorus never sounded better.

"Do you allow women?" we asked.

"Have to. Equal rights."

"Good God, man, ladies can't chew tobacco. They'd have to spit."

"No way. We have a special sniffing department for the ladies."

We were alarmed. "What do they sniff?"

"Snuff. It's the latest thing. Beats cigars, and it's very ladylike. All the kings and queens did it. Snuff's for royalty, man. No smoke, no spitting, just sniffing."

It looks to us as if we've come a long way backward. All progress is change, but all change ain't necessarily progress.

Ashes to Ashes

> It is deplorable to think of a parish where there are 30,000 people living
> without a Christian burial.
> *English Clergyman, 1850*

Kimberly Manning kept her mother's ashes in a paper bag which was now inside a cardboard box ready to move to her new home.

When her husband went back home for the last time to retrieve the last of their belongings — *horrors!* The box was gone. The only thing left was a jar of birdseed, no mama.

Associated Press has reported this tragic event. It seems that the employees with the complex where Kimberly lived had tossed out all the remaining goods. Kimberly had planned on scattering her mama in the Grand Canyon. Now she fears her mother is in the county dump. We can visualize something like this happening next:

Kimberly goes to the dump in search of her mother. There she meets a vagrant rummaging through the trash.

"Have you seen my mother?"

"I dunno. What did she look like?"

"She's in a brown paper bag."

"Come again, lady? You must be kidding."

Kimberly explains the tragedy.

"Oh, I see now," says the vagrant. "What kind of bag?"

"Piggly Wiggly."

"Oh my God, I did see her yesterday. I thought she was powdered milk. I tried some in my tea, but it didn't work right. Too much calcium so I gave it to a friend."

"Where is he?"

"Right over there on the other side of the dump."

Kimberly accosts the friend.

"Where's my mom?"

"I haven't seen nobody, lady."

"What about a Piggly Wiggly sack?"

"If you mean that bag full of powdered sugar, I tried some in my lemonade. It didn't taste right so I figured it was flour instead. I gave it to the bakery up the road."

Kimberly rushes to the bakery.

"Is my mother still here?"

"Who is your mother?"

"She's in a brown paper bag from Piggly Wiggly."

The baker shakes his head sadly.

"No wonder. I couldn't figure out why my rolls didn't rise. I figured it must be cement. Here, see for yourself." He gathers up a handful of rolls. "They're hard as a rock. I was going to use them in my garden."

"Oh, Lordy, no, don't do that. I want my mama."

The baker gathers up a full carton of the rolls. "Well, here she is. I hope you can use her some place."

Kimberly goes to the Grand Canyon in a Brinks armored car for security, then flies over the canyon and drops her mother's rolls up and down the river.

Now at last her mother can rest in peace. Ashes to ashes. Rolls to rolls. Dust to dust.

Just the other day the press reported a similar catastrophe. A lady received the wrong ashes; they were not her mother. We're not sure how she could tell, but she sued and won $33 million. Not a bad day's work for some lawyer. We hope her mama got to Heaven in spite of the mix-up.

But as Mark Twain said, "You go to Heaven if you want to; I'd rather stay here."

Bubbas

> A bubba is anything that's large, laid back, and loveable.
> *Bubba Bechtol*
> *Bubba-in-Chief*

Bubbas are a southern male institution, part of the culture. Generally a bubba owns a hound dog, a new gun, an old pickup, a genuine Confederate flag, drinks beer, and eats hot dogs. Although President Clinton owns a dog, he is not qualified as a bubba. Bubbas do not jog, play golf, play the saxophone, or marry lawyers.

There are no female bubbas, and there is no such thing as a Yankee bubba. The only thing comparable to a bubba up north would be a "junior." But a junior can't hold a candle to a bubba.

We are very fortunate locally that the world's chief bubba, Bubba Terryl Bechtol, lives right here among us. He travels the lecture circuit explaining what a bubba is and is not. Bubbas are an unreconstructed southern tradition, he says.

We recently read several news stories so unique that they must have involved bubbas for sure. No other creatures could have pulled off these events.

In Elba, Alabama, a towing company was called to tow away a parking violator. The company arrived, hooked up, and towed off the offending car with an 85 year old grandmother still inside. They later discovered the poor lady in the back seat.

We learned that Bubba's Beer, Bait, and Towing Co. was the culprit. We later interviewed Bubba himself.

"How did this terrible thing happen?" we asked.

"Gosh, I don't know. I seen what looked like some lumps in the back seat, but I never dreamed it was an old lady. But we give her a beer and some fresh bait, and she was OK."

Another bubba news event occurred when a roofing company tore the roof off the wrong house. Needless to say, the owner was irate. We talked to the proprietor of Bubba's Anteeks and Roofing Co. who performed the work.

"How could you make such a dumb mistake?" we asked.

"It was easy. It wasn't really a big mistake. The right house was right next door. We were close. Last week we did the same thing, but those two houses were two miles apart. Boy, was that guy mad! He wouldn't even take a free hot dog. All this second guy had to do to get his new roof was move next door."

The worst bubba tale we ever heard is not fit for polite company so this story is for bubbas only.

It seems that a bubba who worked in a brewery fell into a vat of beer and drowned. His widow was naturally distraught. "Poor Bubba," she wailed, "I guess he never had a second chance."

A fellow worked tried to console the widow. "Oh, yes ma'am, he did. He got out three times to go to the bathroom."

Nobody but a bubba could get away with such a thing. Nothing else on earth compares to a bubba.

Nothing.

Old John Barleycorn

> Water taken in moderation cannot hurt anybody.
> *Mark Twain*

Eventually whiskey will get you into trouble unless drunk in moderation. Drunks are pathetic souls but often very humorous unless they get out on the highways.

Recent medical studies show that the younger you start drinking, the more apt you are to become an alcoholic. Schools and whiskey don't mix.

To his sorrow one poor eighth grader found that out recently when his father wrapped up a bottle of Bordeaux wine and sent it to school with junior as a gift to his son's French teacher. Sound innocent enough? Hardly. The school suspended the boy for ten days for violating a rule about bringing alcohol onto school property. It is not as if the boy threw a party for his classmates and drank the wine, but no matter. Ernest Hemingway helpfully pointed out that "No, a Bordeaux is not a house of ill repute." Small consolation. Alas. The boy now has a record.

Even animals are not exempt from the evils of John Barleycorn. Recently the press reported that a herd of African elephants got drunk (God knows how) and ate up a native village on a drunken rampage.

You will remember the flocks of birds which landed in Iowa and got drunk on some forbidden fruit. The birds ended up doing loops and slowrolls and other acrobatic tricks. Some passed out. They are now attending AA.

Although a drunk is a pitiful specimen, hundreds of funny drunk stories abound; they must be recorded for posterity. These particular drunks are not the dangerous kind.

There were three drunks chasing a train in an attempt to board it. Two of them managed to catch it and climb on, but the third one missed. The stationmaster told the man he was sorry he missed his train. "You don't know the half of it," complained the drunk. "Those other two guys just came down to see me off."

Another drunk was rushing to catch a ferry. He jumped at the last minute and barely made it onto the deck. He injured himself in the attempt. "Whew, I just made it," he exulted.

"Yeh, but what's your big hurry?" asked the deckhand. "We were just coming in to dock."

A master-of-ceremonies pounded his gavel trying to get order. By mistake he hit the inebriated gentleman who sat next to him on the top of his bald head. "Please hit me again, harder," said the man, "I can still hear the speaker."

One mayor reported that his town was so small that the citizens had to take turns acting as the town drunk.

Elsewhere, voters go to the polls to vote wet or dry. William Allen White, editor of the Emporia Gazette wrote that "Kansans will vote dry so long as they can stagger to the polls."

Here's to good old John Barleycorn. But not on the road.

English 101

> If English was good enough for Jesus Christ, it's good enough for me.
> *A Congressman arguing against*
> *a multi-lingual America*

Said the radio weatherman, "The tide is one feet."

We constantly hear the King's English massacred in speech and in print every day even by those who should know better. We don't know how they're teaching English in school these days, but it doesn't always seem to come out right in public.

As George Bernard Shaw once proclaimed, "England and America are two countries separated by a common language." The British brought English to America centuries ago, and we are still trying to kill it. Correct is still correct in 1996.

If you watch C-Span's "Questions to the Prime Minister" on TV, you will hear perfect English spoken by the members of parliament with no stuttering, no ah's, er's, or and's which we Americans throw in to fill up empty space. Even the man on the street in England beats us.

For example, the other day we heard "me or him is going to the police." This is a capital offense. No Englishman would be caught dead uttering such an atrocity. But there are lesser evils. Test your own English as follows:

Do you say "to boldly go" or "to go boldly?" The latter is correct. We still prefer not to split infinitives, but it is done everyday in polite society.

We hear "Each one gets *their* own way." Wrong. "Each one gets *his* own way" or "*his* or *her* own way." "They all get *their* own way" is also correct.

"That is the *reason why* we are going." Wrong. "That is the *reason* we are going" or "that is *why* we are going," but not both *reason* and *why.*

"The dedication and the integrity of the candidate *is* reassuring." No way! "Dedication and integrity *are* reassuring." You wouldn't say "The dog and the cat is fighting." We hear and see this crime committed daily.

"My way is different *than* his way." Wrong. "Different *from*" is correct.

We hear "fiscal" pronounced "physical" even by lawyers. Fiscal means money, not pushups.

"*Who* are you giving that to?" Wrong. "*Whom* are you giving that to?" Pure purists won't even end a sentence with the preposition "to." One of Winston Churchill's secretaries corrected him for this indiscretion, and he exploded, "That is one inconvenience up with which I shall not put." Take your choice.

"*Who*" and "*whom*" can confuse you if you're not careful. "*Who*" is a subject, "*whom*" is an object.

We hear "It's not *me*" all the time. Incorrect. "It is not *I*" or "I am not it" but never *me* in this case and certainly not "It ain't me."

"If I *was* you" is second-class English. "If I *were* you" is correct.

"It's a case of *him* doing wrong." Wrong. "It's a case of *his* doing wrong."

"Do *like* I do, not *like* I say." Wrong. "He looks *like* the devil" is correct, but "Do *as* I do, not *as* I say."

And there are those who insist on saying "At this point in time." They mean "Now."

And so on. A few of the rules of grammar are complicated, but most are not. Correct English is like correct arithmetic: $2 + 2 = 4$, not 5.

We can all do better. The American Revolution ended in 1783. Let's not keep on shooting the King's English. The King is still correct.

Until Death Do Us Apart

> I was married by a judge. I should have asked for a jury.
> *Groucho Marx*

Divorce is tragic. It breaks up a family, and sociologists blame broken families and single parenthood as the root cause of lots of our current social problems. Unfortunately, the divorce rate is headed for 50%. Marriage is no longer permanent.

Fortunately, if there is a bright side to divorce, some of those who have suffered this disruption can still joke about it. We offer some expert advise and opinions from those who should know:

> My husband demanded a hot breakfast. I set his corn flakes on fire.
> *Divorcee*

> I am a marvelous housekeeper. Every time I leave a man, I keep his house.
> *Zsa Zsa Gabor*

> Bigamy is having one husband too many. Monogamy is the same.
> *Attributed to Liz Taylor*

> A lot of people have asked me how short I am. Since my last divorce I'm about $100,000 short.
> *Mickey Rooney*

> Marriage is the chief cause of divorce.
> *Groucho Marx.*

> I was happily married for 15 years. 15 out of 25 ain't bad.
> *Divorced husband.*

> I was married so many times my face is pitted with rice.
> *Hollywood actress.*

> Marry in haste; repent at leisure.
> *Shotgun husband.*

Then there are great divorce stories. A husband filed for divorce. The judge asked him why. "Well," he said, "she talked too much."

"What did she talk about?" asked the judge.

Said the husband, "She never did say."

During a sermon on marriage, the preacher said to the congregation, "Will those of you who think you're perfect please stand up." One man stood up in the rear of the church.

"Do you really think you're perfect?" asked the preacher.

"No, sir," said the man, "I'm just standing up for my wife's ex-husband."

An old Italian proverb says that "After the ship has sunk, everyone knows how she might have been saved." Before the ship of marriage goes down, we offer some marriage counseling borrowed from a *Great Quotation's* booklet entitled "Things you'll learn if you live long enough...so you might as well know now:"*

"The person who marries for money usually earns every penny of it."

"If men acted after marriage as they do during courtship, there would be fewer divorces and more bankruptcies."

"Behind every successful man stands a surprised mother-in-law."

"When it comes to broken marriages, most husbands will split the blame half his wife's fault, and half her mother's."

"Marriages are made in Heaven so are thunder and lightning."

"Marriage is like twirling a baton, turning hand springs, or eating with chopsticks. It looks easy till you try it."

Married persons are barely batting 50% longevity. That means 1½ strikes and you're out.

* Used with permission and published by Successories, Inc. All rights reserved.

Outhouses Are Back In

> Laughing at our own mistakes can lengthen our own life.
> Laughing at someone else's can shorten it.
> *Cullen Hightower*

Gulf Breezians still hear a giant flushing sound as the Escambia County Utility Authority tries to wash the brown stuff out of their water pipes.

Last week housewife Fannie Flushwell, in a fit of despair, recommended going back to outhouses to cut down on water and to get even with ECUA. History has shown that we survived for centuries without indoor plumbing. What to do?

For some expert advice we found inventor, computer genius, and plumber Frank Hobbgood tinkering with his new and improved space commode which will retail for $19.95.

"Should we go back to outdoor plumbing?" we asked him.

"You're danged right. Commodes don't hack it. 1.6 gallons don't do it. I tried to tell'em up in Washington, but they didn't pay no attention. I'm a 12 to 15 quart man myself; that's what it takes to do a proper job. But that's just wasting more water. Outhouses don't use no water. No plumbing, no machinery, no nuthin'. Simple! Beautiful!"

"What if some mean boys tip them over?" we wondered.

"Not too serious unless you're in it. But we can tie'em down good, put on an extra-heavy anchor just like the Navy. They ain't going no place. They're the answer, no doubt about it."

We can just picture the beauty of this thing. Dig a hole, prop up a neat little 4'x4'x6' bungalow on top, put in a Sears catalog, and you're in business. With a half-moon cut in the door, no one can possibly mistake it for a phone booth. You naturally don't want any strangers rushing in to make phone calls while you're seated.

Outhouses are mechanically fool-proof and environmentally correct. They can be built of wood, tin, asbestos, fiberglass, plastic, or even brick. They can be camouflaged with paint to blend in with nature or else hidden behind a bush. No chains to pull, no handles to fool with, no wasted water.

There are many assorted models to choose from: single cockpit, dual cockpit, 2-holers on a 3-hole chassis, etc. The ultimate luxury model is a brick colonial 2-story job with wall-to-wall carpet, stereo, and TV. This model is armed with a small cannon to fight off attackers. It is non-tippable to prevent looting, and cannot be carried off by pranksters to put in the preacher's front yard as we used to do. Pure peace and quiet and security. Our ancestors did it; why don't we?

The outhouse solution completely eliminates the need in Gulf Breeze for any more brown water flushes; 1.6 gallons aren't enough, anyway. After all, the portable potties you see dotting the landscape all over town certainly show that our citizens are accustomed to such facilities. No further detailed instructions would be necessary.

Then, if ECUA can't plug up the brown water, Gulf Breezians can drink bottled water and can wash in the bay, thus eliminating any more need for ECUA water, period! That will take care of that.

Hallelujah!

Burn, Baby, Burn

> It is impossible to underestimate the intelligence of the American public.
> *H.L. Mencken*

The Indians had it right: "Don't criticize someone until you've walked a mile in his moccasins." Then when you do criticize someone, you'll be a mile away, and you'll have his moccasins.

We hate to be critical, but humans are now defacing themselves by all manner of weird practices.

First we read the pitiful tale of a young mother who had herself tattooed from head to foot. Her mother had warned her not to needle herself permanently with indelible ink, but she said she didn't care and inked herself anyway with various murals and graffiti.

Now she cares. She has a baby to explain this discoloration to. She is ashamed to appear in public unless wrapped in a blanket or tent or poncho or some other wrapping. Not a pretty picture. It's hard to scrape off a tattoo and also expensive if not painful.

That's just part of her problem. She also had herself pierced. That's another fad gone wild. You see supposedly human beings walking around who have speared every imaginable part of their anatomy, and some unimaginable: ears, noses, lips, tongues, belly buttons, and much worse. Pins and jewelry sticking out all over them. If you repent, can you plug up all the holes?

Just what all this proves is beyond fathoming. We think of cavemen and tribesmen spearing their enemies, but they certainly never speared themselves. They were smarter than that. They stuck their spears into somebody else; it's less painful that way.

Now the latest fad is branding. *Branding*! Why in God's name would any human permit him or herself to be branded with a hot poker? This is just one step from self-immolation.

When we think of branding, we normally think of cattle; but we never saw a cow dumb enough to lie still and get branded if it could help it. As any branding victim should do, the cow moos and bellows and hollers bloody murder.

In order to brand cattle, the cowboys rope them and wrestle them to the ground. Then they tie them and sit on them while the head torturer administers the hot iron to their rumps. Brand X or Circle M or Bar K, whatever, just so the rustlers and sheriffs know who owns each cow. Then you know which rustler to hang for stealing which cow. There is a practical purpose here.

Now come the humans. Do they bellow and holler and fight for their dear lives? Not a chance, according to a victim recently branded. He loved every minute of it, he said. He just lay down and enjoyed the whole affair.

Imagine lying there at ease while the head ghoul fires up a branding iron to 1500°F. He puts a little antiseptic on the chosen spot and *whammo*! He lays that 1500° poker on you for up to 3 seconds while your flesh sizzles medium rare. Did the victim howl? No. Scream? No. He enjoyed it, said it gave him a lift.

What does the brand do to you? It burns a nice piece of flesh the size and pattern of the brand. You can have whatever marking you want to distinguish you from the other cows. No one knows whether the scar will turn out red, white, or blue, or all three. You just take potluck.

Good Lord, imagine submitting yourself to torture voluntarily without a whimper. Chinese water torture would be much more enjoyable. Any such treatment violates the Geneva Convention against mistreating prisoners.

The apes, cavemen, and our other ancestors never dreamed their heirs would grow up to attack themselves. The Indians were smart enough to put on war paint, not branding irons and tattoos. Paint rubs off with turpentine. The Indians pierced Custer and others, not themselves, with arrows.

Tattoos, piercing, branding. What next? We vote for the cows. They're not so dumb. At least they don't get tattooed.

ON GOVERNMENT

Government can't give you anything it didn't take away from you
in the first place.
Senator Barry Goldwater

There are two things which ought never to be seen in the making—
sausages and laws.
German Chancellor Otto Von Bismarck

Next to being shot at and missed, nothing is as satisfying as a tax refund.
Anonymous

The Eiffel Tower is the Empire State Building after taxes.
Anonymous

You can trust the government. Just ask any Indian.
Bumper sticker

What Goes Up Comes Down

The doctors X-rayed my head and found nothing.
Pitcher Dizzy Dean
(after being hit on the head with a ball)

We continue our series on government bureaucracy, forms, rules, and regulations. Dot Brown years ago printed this classic example of bureaucratic paperwork in her column in the *Pensacola News Journal*. Quote:

"I am writing in response to your request for additional information.

"In Block 3 of the Accident Report Form, I wrote 'trying to do the job alone' as the cause of my accident. You said in your letter that I should explain more fully, and I trust that the following details will be sufficient.

"I am a bricklayer by trade. On the date of the accident I was working alone on the roof of a six-story building. When I finished my work I discovered that I had about 500 pounds of bricks left over. Rather than carry the bricks down by hand, I decided to lower them in a barrel by using a pulley attached to the side of the building at the sixth floor.

"Securing the rope at ground level, I went up to the roof, swung the barrel out and loaded the bricks. Then I went back to the ground and untied the rope, holding it tightly to ensure a slow descent of the barrel full of 500 pounds of bricks.

"You will note that in Block 11 of the report form that I weight 135 pounds.

"Due to my surprise at being jerked off the ground so suddenly, I lost my presence of mind and forgot to let go of the rope.

"Needless to say, I proceeded at a rather rapid rate of speed up the side of the building.

"In the vicinity of the third floor, I met the barrel coming down. This explains my fractured skull and collarbone.

"Slowed only slightly, I continued my rapid ascent and did not stop until the fingers of my right hand were two knuckles deep in the pulley.

"At this time I regained my presence of mind and held on to the rope in spite of my pain. Moments later, the barrel hit the ground. The bottom fell out of the barrel. Devoid of the bricks, the barrel weighed about 50 pounds.

"I refer you again to the information in Block 11.

"As you might imagine, I began a rapid descent down the side of the building.

"About the third floor, I met the barrel coming up.

"This accounts for the fracture of both ankles and the lacerations of my legs and lower body.

"The encounter with the barrel slowed me enough to lessen my injuries when I fell onto the brick pile, so only three vertebras were fractured.

"I am sorry to report, however, that as I lay there on the bricks in pain, unable to move, and watching the empty barrel six stories above me, that I again lost my presence of mind and let go of the rope.

"The empty barrel came down, and broke both of my legs.

"I hope I have furnished the information you required about the cause of the accident."

Unquote!

Red Tape

> Bureaucracy is a giant mechanism operated by pygmies.
> *Honoré de Balzac*

Congress has declared war on bureaucrats in Washington, D.C. Congress wants to cut costs, and that means some bureaucrats will have to go.

Bureaucrats are the poor nameless little souls wearing green eyeshades and glasses who sit hunched up over their desks writing regulations, inventing paperwork, or poring over someone else's paperwork. They are the government, and they are not there to help you.

Some definitions will help you better understand how government bureaucracy works:

Expert - An "ex" is a has-been and a "spurt" is a drip under pressure.
Colleague - Someone called in at the last minute to share the blame.
Program - Any assignment which can't be completed by one phone call.
Implement a Program - Hire more people and expand the office.
Coordinator - A guy who sits between two expediters.
Expedite - To compound confusion with commotion.
Activate - To make more copies and add more names to the memo.
Research - Go looking for the jerk who moved the files.

In order for you as a citizen to do anything today in our democracy, you have to fill out a form and send it in for approval. Somewhere in government there is always someone who was hired to process your request.

Following is one good example of what happens when an expert government coordinator gets hold of your paperwork and reviews the forms which he invented in the first place:

Some time back, Congressman William Lehman of Florida hosted a hearing by a subcommittee of the House Committee on Post Office and Civil Service in Miami to discuss federal regulations and paperwork. One witness was a Louisiana parish (county) official who traveled to Miami to tell his tale of woe.

It seems that a development company was planning a new complex in his parish, and secured the approval of all 23 agencies (local, parish, and state) just to find that they must also apply to the U.S. Department of Housing and Urban Development. The company filled out the appropriate forms and mailed them off to Washington. They received the following letter as acknowledgment from some bureaucrat:

"We received today your letter, enclosing application for your client and supported by abstract of title. We have observed, however, that you have not traced the title previous to 1803, and before final approval, it will be necessary that the title be traced previously to that year."

Stunned by the response, the developer's attorney sent back the following:

"Gentlemen: Your letter regarding the title received. I note that you wish title to be claimed further than I have done it.

"I was unaware that any educated man failed to know that Louisiana was purchased from France in 1803. The title of the land was acquired by France by right of conquest from Spain.

"The land came into possession of Spain in 1492 by right of discovery by an Italian sailor named Christopher Columbus. The good queen Isabella took the precaution of securing the blessing of the Pope of Rome upon Columbus' voyage before she sold her jewels to help him.

"Now, the Pope is the emissary of Jesus Christ, Son of God. And God made the world. Therefore, I believe that it is safe to assume that He also made that part of the U.S. called Louisiana, and *I hope to hell you're satisfied.*"

Go Seminoles

I wonder what's the matter with those Indians.
They seemed to be all right at the dance last night.
General George Armstrong Custer at the Battle of Little Big Horn

We are not talking about Florida State University Seminoles here. We're talking about *real* Seminole Indians.

Recently the state of Florida stopped all persons operating bungee jumping in the state. The Seminole Indians went right on jumping in spite of the ban. This disregard of a state mandate infuriated the state administration, so Governor Chiles called in his attorney general for consultation.

It turns out that the Seminole Indians never surrendered to the United States. They are, therefore, a foreign nation, the Free State of Seminole. They are independent, just like the Vatican or Great Britain or any other foreign country.

"Good God, how can this be?" asked the perplexed governor.

"It says so right here in the minutes of the campfire meeting," said the attorney general.

"This will never do. Go negotiate a treaty," ordered the governor.

The attorney general set out to visit Seminole State but was stopped at the border and sent back to get a visa.

On his return trip he got an appointment to meet with Chief Wigwam. The attorney general presented his official state credentials and also the regulation against bungee jumping.

"You go jump!" said the chief. The Indians then dropped the attorney general off the tower. He bounced up and down like a Yo-Yo. The braves left him dangling upside down so that he might consider modifying his attitude.

When word of this indignity got back to Tallahassee, the governor was outraged. He called the U.S. Secretary of State for help.

"Let's declare war," he pleaded.

"Not yet," said the secretary. "We'll try a treaty one more time."

The secretary then sent his deputy to Seminole State with a truckload of trinkets. Chief Wigwam was not at all impressed.

"This no Manhattan Island," said the chief. "Me no want no damned beads. Me want wampum." The deputy went back to Washington, D.C., to report.

"All right," said the secretary. "They're eligible for foreign aid. But in return they'll have to comply with all the federal regulations: environment, HRS, welfare, drugs, growth management, OSHA, IRS, bungee, bingo, racing, asbestos removal, alligator wrestling, etc."

The deputy returned to Seminole State with a truckload of federal regulations. He relayed the secretary's message to Chief Wigwam.

"First we smoke peacepipe," said the chief. He lit a fire in his pipe. Then he tied the deputy to a stake surrounded by piles of federal regulations and lit another fire.

Things were getting out of hand. The president sent his secretary of defense with instructions either to sue for peace or declare war. The chief did not take kindly to these ultimatums. The Indians held a tribal council meeting to vote followed by a powwow complete with feathers, war paint, and war dances. The U.S. delegates were found guilty and taken prisoner. Some were condemned to play bingo for life.

The remaining delegates were then freed and returned to Washington, D.C., without their scalps as this is still a popular old Indian custom. The delegates were last seen trying to buy their scalps back from Yankee tourists.

To this day the Seminoles continue to bungee jump, scalp bureaucrats, play bingo, wrestle alligators, have fun, and bring in the tourists with wampum.

It is a very comforting thought to Chief Wigwam to be eligible for foreign aid without having to comply with all the domestic rules and regulations which go along with being a non-Seminole. It is also comforting to know that the Seminoles can paddle their own canoes. They're just smarter than the rest of us.

So put that in your peacepipe and smoke it!

Seminoles Uprise Again

> I wonder who greased this grapevine.
> *Tarzan*

The Seminoles (real Indians, not the Florida State University team) are on the warpath again.

You may remember that just recently the Seminoles rebelled against the state's anti-bungee jumping injunction and scalped several dozen bureaucrats to prove their point. The scalpings were pretty convincing. It is extremely dangerous to mess around with people with tomahawks.

Now the Seminoles have petitioned the state to add high wager, Las Vegas-style gambling to their bingo and bungee operations. A federal law provides that Indians on their reservations have a free hand to govern themselves as they see fit. They deserve this right based on what we did to the Indians and the bison.

The attorney general of Florida challenges this view with a legal opinion that the Seminoles can do anything that's legal, but casino gambling is not legal in Florida. He had better get a good grip on his scalp.

Needless to say, Seminole Chief Wigwam is furious.....again. The government insists on continuing to aggravate him and his braves.

We caught up with the chief watching the state attorney general dangle from a bungee cord. The chief was also filling out his bingo card at the same time. We asked the chief what about this casino gambling?

"Um gallo gallo," thundered the chief. We're afraid this response would be unfit to print if translated.

After the chief recovered his composure, he continued, "State of Florida have bingo, lotteries, jai alai, horse racing, and dog racing, and now them try tell us we no gamble. Humpf! Um gallo gallo." He brandished his tomahawk for emphasis. We quickly got the point he was making. We wondered why the state attorney general didn't catch on.

"What does your own attorney general say about this?" we asked the chief.

"Him better agree with me!" warned the chief.

"Well, what on earth can your tribe do about it? The law is the law."

"Yesterday on earth we capture the state attorney general," said the chief proudly. "Right now we adjusting his attitude for negotiations."

We could see the state attorney general still bouncing up and down on his bungee cord. "When will you negotiate?" we asked.

"After him stop bouncing," answered the chief with glee. "Then we fight duel to see who right."

"Good God, chief, you can't do that. It's not legal," we warned.

"It legal for Seminoles," he said.

"How come?" we asked.

"Because me say so," said the chief. His attorney general quickly nodded his head in agreement.

"Will the state attorney general have his choice of weapons?" we gasped.

"Him want to play 5 card stud for duel," said the chief with a disgusted look. "That no deal. It no legal in Florida," he advised, proud of his legal opinion. His attorney general quickly agreed.

"What *is* the choice of weapons?" We were hesitant to ask.

"Me choose bow and arrow," replied the chief with satisfaction as he sharpened his arrows.

We were dumbfounded at this sad turn of events. Just then the governor and lieutenant governor of Florida arrived to serve as seconds for their attorney general. We couldn't stand the thought of our state attorney general shot full of holes.

As we were leaving, the Seminoles cut the state attorney general down from his bungee jump and prepared to give him one practice round with the bow and arrow which the Seminoles had rented him. The chief had already taken ten practice rounds. Every arrow was in the bull's-eye.

We could see the governor and lieutenant governor down on their knees begging for mercy for their attorney general. Both were sobbing audibly.

"Um gallo gallo," we heard the chief say. We don't think this was a very nice thing to say to our governor.

The outcome of this event is too awful to imagine. We still don't know how it all turned out. You'll just have to read about the ending in your newspaper. We can't even bear to think about it.

Remember General Custer?

The Second Coming

> It gets late early out there.
> *Yogi Berra*

(Editor's Note: The following dissertation involves a local sales tax election and what to do with the money.)

The sales tax question is upon us again! We will vote on whether to raise our sales tax one cent to solve some big problems and make some community improvements. The alternatives would seem to be to raise ad valorem (property) taxes, or to do nothing, keep our money, and keep our problems. What will it be?

Tourists will pay 38% of the sales tax. They use our public facilities but don't otherwise help pay for them. Only property owners with values over $25,000 pay ad valorem taxes; there are a lot fewer of them than there are tourists.

The City and County have published a list of projects to be funded over 10 years by this extra one-cent sales tax. Included in these projects is the extension of the I-110 interstate spur north of I-10. We need this improvement because the North Davis Highway area has become a traffic battle zone.

We took a trip out that way just to see for ourselves. First we drove out on I-110 and tried to get off on Davis Highway. Traffic was stalled bumper to bumper. After several days' wait, we backed up to Fairfield Drive and started over. We tried going out Davis Highway, but there was a two-month delay there from traffic light to traffic light (when one turns green, the next one turns red at the same time). Traffic was backed up and stacked up. We finally decided that you can't get there from here.

We never say die, so we hired the Ellis Davis Commuter Airline Company and flew out and landed on the grass strip on the interchange cleared by the Clean Community Commission Saturday Morning Clearing Brigade. They have cleared a large area, but there still remains a deep, dark forest. People were camping back in there. We interviewed several to get our story.

The first person we talked to was Don Tristan De Luna Junior, a direct descendant of Don Tristan De Luna Senior, the Spanish explorer who discovered Pensacola, Florida. The De Luna family has been camped there since 1559. "How come?" we asked.

Don Tristan Junior said sadly, "Don Tristan Senior planned the I-110 extension in 1561. He applied to Tallahassee and Washington, D.C., for funds and never heard back. A sales tax referendum failed in 1562, and we've been waiting ever since."

We couldn't quite believe all this. "Did you quit trying?" we asked.

"Don Tristan Senior finally got discouraged and flew back home. We just keep hoping for the extension someday so we can get out of these woods and go back to Spain."

We reported this finding to historian John Appleyard, who has been hunting for the De Lunas since 1957.

We then moved on a little farther to the next homestead—a log cabin. This family came overland from Minnesota in 1874 and got caught in a traffic jam at Olive Road and Davis. They camped overnight and are still here waiting for the traffic light to turn green.

We next interviewed Kermit O'Brien and his family who emigrated from Ireland in 1919. I asked Kermit what the problem was.

"Our visa ran out and we had to apply for another one. We got it in 1927. We tried to leave several times, but we just got caught in traffic and couldn't move. Besides, we like it here. The highway department cuts our grass!"

We talked to several tourists from New England who were camped in their cars in traffic. They all handed us their wallets and credit cards and held their arms up in the air. "What's this for?" we asked. "This is not a holdup."

"Take that and more. That's all we've got, but we would gladly pay sales tax if we could just get to the beach. We started in April, and it's already August."

One man stalled in traffic looked desperate and asked if we could help. "The five year warranty on my car is running out, and I need to get to a garage. Will we be able to get out of here anytime soon?" he asked.

We said we didn't think so without some sales tax money and reminded him to vote. He said he would be glad to if he could just see his wife and children one more time.

Our last interview was with a political candidate who started running for office in 1984. He has been campaigning at the corner of Olive and Davis for six years. "I can't even get home to vote," he said. "Was I elected?"

We sincerely hope that it doesn't take the second coming of Don Tristan De Luna Senior to get us out of this mess. We're still stuck here in traffic at Olive and Davis (we telegraphed our report to the paper).

(P.S. The tax passed. The road has not been fixed. Those people are still stuck out there.)

Privy Progress

> There is nothing so wasteful as doing with great efficiency that
> which does not need to be done at all.
> *Anonymous*

Trouble in paradise: our national parks. Congress is up in arms ... again.

The National Park Service under Secretary Bruce Babbitt spent $584,000 per house to build 23 single-family, 1800 square foot residences at the Grand Canyon for use by park employees. That's some of your money. Think about what you could build on your lot for $584,000. That kind of money sounds more like La Jolla, California, than a wilderness park.

On top of that miscarriage of government, try this on for size: In a remote area of wilderness in Delaware Water Gap, Pennsylvania, the park service built a $330,000 state-of-the-art composting outhouse with slate roof and customized exterior paint at $78 a gallon. The outhouse features a 29-inch thick foundation, composting toilets, and landscaping with wildflower seeds costing $720 a pound. One-third of the cost was overhead, including architects and engineers, according to the Associated Press.

How does an outhouse compost?

We had to find out for ourselves so for some reliable answers we went to Delaware Water Gap to visit this deluxe composting privy. You never know when you might need one. We talked to Fuller Flusch, the park ranger in charge of outhouses.

"Tell us about the 29-inch thick foundation. That would hold up the Empire State Building," we said.

"Exactly. We may want to add 50 or 60 floors eventually. Twenty-nine inches will do it. During wartime we may have to drive a tank in there. This will hold up several Sherman tanks."

"We see lots of wildflowers growing naturally outside here. What about those wildflower seeds you planted for $720 a pound?"

"Those were authorized, government-issue seeds, grown, approved, tested, and inspected by the FDA. You wouldn't want just any old unapproved weeds growing around a $330,000 outhouse."

"That gets us to the inside of your privy. What ran the price up so high?"

"This is no ordinary outhouse. We feature individualized stalls for comfort and privacy. Our people used to just run out and get behind a bush. That won't do."

"What's in these stalls?"

"I thought you'd never ask. They're only 1000 square feet, but they're loaded. You get a radio, TV, fax, computer, phone, wet bar, recliner, four-poster bed, desk and chair, library, wall-to-wall carpet, and an original Picasso."

"Is there actually a potty in there with all that other junk?"

"I'm glad you asked. That's the whole purpose of the building. We feature customized, composting, fur-lined potties with padded seats using thick foam rubber cushions covered with French tapestry from Chateau de la Creme. They're a delight."

"My God — it sounds more like Queen Elizabeth's throne. You might fall asleep."

"Got that covered. The valet will wake you up gently and put you to bed. You don't have to lift a finger. Even the potty lids are hydraulically operated up and down. No strain."

"What about this composting? Couldn't you just bring a few bulls in here a lot cheaper?"

"Don't be crude. Manure is too bourgeoisie."

"What does Congress say about all this?"

"They're mad as hornets."

"What are they going to do about it?"

"They're moving Secretary Babbitt from Washington to stall #1."

"What will he do out here?"

"Compost."

Bureaucracy Strikes Again

> We're lucky we don't get all the government we pay for.
> *Dr. Milton Friedman*

Pedro Gonzalez wants in. This is a true story.

Immigration is hot these days. Lots of folks outside the U.S. want to come in. Lots of folks inside don't want any more folks from the outside, period. They think we're already too crowded.

California, Florida, and Texas are all taking hard looks at cutting down on immigrants, legal and illegal.

The laws put strict limits on immigrants, so many each year. Immigration laws are tough and strictly enforced, so strictly that we recently heard the sad but true story about Pedro (whose name has been changed to protect the bureaucracy).

It seems that Pedro has only seven fingers; three are missing. Pedro reported to the immigration service for processing. He was met by bureaucrat Emma Grant who took his fingerprints.

"Wait a minute here," said Emma. I've only got seven prints. I need ten."

"I don't got ten," advised Pedro. "I only got seven."

"Where are the other three?"

"I lost 'em in a sawmill. I no got any more."

"This will never do. You can't immigrate with seven fingerprints. We have to have ten. Go find three more someplace."

"Where I gonna find?"

"Beats me. I'll send you to my supervisor." It turns out her supervisor is her husband, M.A. Grant.

"What seems to be the trouble here?" asks Grant.

"I no got ten fingers," says Pedro.

"You foreigners are all alike. You don't follow instructions. The law is clear on fingers. You have to have ten fingerprints."

"Where I find?"

"I don't know, but we can't let you in with only seven fingerprints."

"How about my wife? Can I use three of hers?"

"Well, that's an idea, but then she'll only have seven left, and we can't let her in. You do want your wife in too, don't you?"

"Oh, Lordy, what will I do? I need my fingers, and I need my wife bad. How about some toes?"

"Toes don't count. I recommend you see your psychiatrist. Maybe he can help."

Pedro calls his psychiatrist, Doctor J. Throckmorton Quackinbush.

"What can I do for you?" asks Quackinbush who is on his cellular phone on the way to the golf course.

"I need three more fingers," begs Pedro.

"Well, take two aspirins and see me in six months," advised the doctor. Pedro visits the doctor six months later.

"Hold up ten fingers," says Dr. Quackinbush.

"I no got ten. I only got seven."

"Man, think positive."

"I thinkin' positive, but I still only got seven."

"Look here, man, if you don't cooperate, I can't help you. You had better call your congressman."

Pedro calls the Honorable Henderson Votewright, M.C.

"I needa law with only seven fingerprints in it," pleads Pedro.

"But we always use ten," reminds Votewright.

"But I no got. I only got seven."

"Well, all right, I'll write a new law and attach it to the abortion bill."

"But I no want no abortion. I just want my fingers."

"That's OK. That's just the way we legislate."

The president vetoes the bill, and Pedro is left unimmigrated with only seven fingerprints. If anybody out there has some extra fingers, maybe you can help..

Water, Water Everywhere (but not a drop to drink)

> Politicians should always be sincere whether they mean it or not.
> *Pat Paulsen*
> *United We Sit Party*

The Gulf Breeze City Council is fighting mad.

It claims the Escambia County Utilities Authority (ECUA) is coloring the water, and Gulf Breeze is declaring war on ECUA. We can just imagine what's going on over there across the pond.

We interviewed Thelma Washtout, a Gulf Breeze housewife.

"What's wrong with your water?" we asked.

"Ugh. It's full of brown stuff."

"What kind of brown stuff?"

"It could be anything, bugs or peanut butter or even sludge."

"Good God, not sewage."

"Well, the rumor is that ECUA is sending us water from the Main Street sewage plant. Whatever it is, we're not going to take it any more."

"Where will you get your water?"

"Anything would be better. If we have to, we'll steal it from the golf course. We're mad as hell."

It certainly sounded like it. Things appear to be worse than we expected. ECUA has issued an ultimatum: we take over your water for 90 days; take it or leave it.

We can just picture what happens next. The Gulf Breeze council does not take this lying down and comes out swinging. Council persons invade the next ECUA meeting in force, headed by the honorable mayor himself.

The ECUA chairman pounds his gavel and calls the meeting to order.

Chairman: We're pleased to have this nice crowd here today.

Mayor: We're not a nice crowd, and we're not pleased to be here. We want you to take back your stinking water.

Chairman: Excuse me. That's an insult. Our sewage plant stinks but not our water.

Mayor: You've got junk in your water.

Chairman: Baloney. Our water tests out pure as the wind-driven snow. Those might just be vitamins in your water.

Mayor: Nuts! We can take vitamin pills if we want them. Besides, we don't take brown vitamins. All we want is water, clean water. We want that brown junk out, or we want our money back.

Chairman: We offered to come over there and take charge, but you turned us down.

Mayor: We don't take your threats kindly, thank you. You need to stay home and fix your water. Just try taking a bath in this stuff. Look at me. I'm turning brown. It's awful.

The mayor shakes his brown fist at the chairman.

Chairman: Nobody shakes a fist at this board.

The ECUA members stand up en masse and shoot their squirt guns full of brown water at the council. The council persons grab a fire hose and squirt the board. The chairman adjourns the meeting as the water rises. They all turn brown and go away mad and soggy, dripping sludge or vitamins or heaven knows what else.

The ECUA takes an axe and cuts the water line to Gulf Breeze in half. Out of water, Gulf Breeze digs its own well and strikes oil. Now there is oil in the water, but nobody is complaining. They are bathing in dollars.

Peace!

Pork

> I always like to see politicians praying. Then I know what they are
> doing with their hands.
> *Bob Hope*

The Clinton administration has announced zero deficit and extra cash to spend in the next federal budget. Oh boy, imagine the glee in Congress. More money to divide up among the robbers.

The Senators have circled their wagons and are preparing to shoot it out for the loot. They are experts, much like bank robber Willie Sutton.

The two top robbers are Senator Byrd of West Virginia and Senator Inouye of Hawaii. Senator Byrd is working on moving Washington, D.C., to West Virginia, buildings and all. Senator Inouye just captured one-half billion dollars for various boondoggles in Hawaii.

How does the Senate divide the spoils? The Senators retire to a large room, lock the doors, and play porker, a game much like poker but played with bare knuckles, minus the Marquis of Queensberry rules. This is serious combat.

The porker games' bank includes the following assets: court houses, federal buildings, aircraft carriers, shipyards, various navy ships, aircraft, veterans hospitals, cemeteries, parks, airports, weather stations, federal judges, military bases, bridges, ports, farm roads, highways, festivals, and national holidays, among other federal handouts.

The game is seven-card draw; Navy bases are wild. In order to open the betting, a porker must hold at least two federal buildings or better. Senator Byrd has no trouble opening.

Senator Byrd: I'll bet the FBI building, a veterans hospital, and two court houses.

Senator Inouye: I'm in. I'll raise you two federal judges and an airport.

Senator Johnson, South Dakota: This game is too rich for me. I fold. (He is in pain and in tears).

Senator Feinstein, California: I'm in for another aircraft carrier, a cemetery, a veterans hospital, a weather station, and the Kumquat Festival. I'm calling you.

Senator Byrd lays down a port, two aircraft carriers, the treasury, a navy base, four army divisions, a dead-end highway, and the Pentagon. No one can beat that hand. There is a lot of moaning.

Senator Inouye, gnashing his teeth: What the hell are you going to do with two aircraft carriers and a port? You don't even have an ocean out there.

Senator Byrd, smiling: The Corps of Engineers is digging it right now.

Senator Grassley, Iowa: I won an army base with two air force wings and no runway. I'll swap the two wings for an army division.

Senator Wellstone, Minnesota: You're on, and I've got a bridge I can't use.

Senator Inouye: I'll take the bridge.

Senator Feinstein, alarmed: Why do you need a bridge?

Senator Inouye: It's always been my dream to build the Inouye Causeway to California.

Senator Feinstein: Good God, I don't believe this. That's pure sausage.

The game ends with some senators gleeful, some shaking their heads, others weeping loudly. They march off to vote pork.

To the victors belong the spoils, or something like that. But as Barry Goldwater put it, "The government can't give you anything it didn't take away from you in the first place.

Amen.

ON HISTORY

History is just one damned thing after another.
Henry Ford

I don't make jokes. I just watch the government and report the facts.
Mark Twain

The race is not always to the swift, nor the battle to the strong,
but that's the way to bet.
Damon Runyon

History teaches us that history teaches us nothing.
T.B. Macaulay

Never Say Die

Never look back. They may be gaining on you.
Satchel Paige

Following up our recent 50th college reunion, we looked back over the last 50 plus years. If you think things haven't changed in 50 years, read on! This is the way we were 50 years ago:

We were before television, before penicillin, the pill, polio shots, antibiotics and frisbees. Before frozen food, nylon, dacron, Xerox, and Kinsey. We were before radar, fluorescent lights, credit cards, and ballpoint pens. For us time-sharing meant together-ness, not computers; a chip meant a piece of wood; hardware meant hardware and soft-ware wasn't even a word. In those days, bunnies were small rabbits, and rabbits were not Volkswagons.

We were before Batman, Rudolph the Red-Nosed Reindeer, and Snoopy. Before DDT and vitamin pills, vodka (in the United States), and the white wine craze; before disposable diapers, jeeps, and the Jefferson nickel; before Scotch tape, M & M's, the au-tomatic shift, and the Lincoln Continental. Before atom bombs, AIDS, open heart sur-gery, organ transplants, Israel, shopping malls, and supermarkets.

When we were in college, pizzas, Cheerios, frozen orange juice, instant coffee, and McDonald's were unheard of. We thought that fast food was something you ate during Lent. We did eat live goldfish.

We were before FM radio, tape recorders, electric typewriters, word processors, Muzak, electronic music, and disco dancing. We were before pantyhose and drip-dry clothes; before ice makers and dishwashers, clothes dryers, freezers, and electric blan-kets. Before men wore long hair and earrings and women wore tuxedos. Bikinis and miniskirts were seen only in burlesque shows. We got married first and then lived to-gether. How quaint can you be? In our day, cigarette smoking was fashionable, grass was mowed, coke was something you drank, and pot was something you cooked in.

We were before coin vending machines, jet planes, helicopters, and inter-state highways. "Made in Japan" meant junk (we were fighting them at the time), and the term "making out" referred to how you did on your exam.

In our time, there were five-and-ten cent stores where you could buy things for five and ten cents. For just one nickel you could make a phone call, buy a Coke, or buy enough stamps to mail one letter and two post cards. You could buy a new Chevy coupe for $600, but who could afford it? Nobody! Besides, gas was rationed along with food, shoes, tires, etc. The draft was a tall beer and also the non-all-volunteer Army. The government volunteered us in those days.

We were not before the difference between the sexes was discovered, but before sex changes. We just made do with what we had.

And so it was 50 years ago ... that is the way we were. Here's to another 50!

Ready, Aim, Fire

> Those who do not remember the past are condemned to repeat it.
> *The Philosopher Santayana*

In 1821 Spain ceded Florida to the U.S. On July 17 we celebrate the 171st anniversary of that great event in Ferdinand Plaza in downtown Pensacola, where General Andrew Jackson stood under an oak tree and Spain handed over Florida.

We wondered what this great event would look like in 1999. Things would have gone something like this:

General Andrew Jackson marched his troops into Ferdinand Plaza at 1 p.m. sharp, following the noon picnic in the plaza. He stood under an oak tree in the shade and prepared to receive Florida from the Spanish governor who was standing out in the hot sun with his staff.

Just as the governor was about to read the official proclamation handing over the key to the Western Gate to the Sunshine State, whereas, therefore, etc., *ad nauseam*, a messenger rushed over from City Hall.

"General, you don't have your parade permit."

"Dammit, man, this is no parade. We're standing still. Besides, the Ku Klux Klan didn't get a permit."

"That doesn't matter. You marched on, you march off. That's a parade. That's when I'll report you. Here, read these regulations."

"By God, I guess you're right." The general sent his wife Rachel, who was selling souvenir dehydrated confederate cannon balls on the sidelines, rushing to City Hall to post a $500 cleanup bond and fill out the forms for a permit. He was also required to have a little man with a broom and a shovel to follow along after the general's horse.

Before Rachel could return, an emissary from the Department of Environmental Regulation (DER) arrived to inquire about an environmental impact study.

"We need to know what all this runoff will do to the U.S."

"The U.S. has been running off on us for years," pleaded the Spanish governor.

"We don't care. This is a new ball game. We don't trust you Spaniards. We want a study. Hire a consultant, or there is no deal. You get me?" The general kindly sent a soldier rowing (the bridge was blocked with traffic) to Gulf Breeze to appeal to DER Headquarters.

"Muchas gracias," said the governor, mopping his brow in the heat.

A state planner rushed up all out of breath. "We can't approve this transfer without a growth management survey. The sewers will have to be able to handle the projected overflow. Here, read these regulations, and then fill out these forms." The forms were loaded in a wheelbarrow standing nearby.

"But we've been handling your (bleep) for years," pleaded the governor, to no avail.

Next in line was the federal pollution control officer, who pointed out several polluted figures lying around under the park benches. More forms. The pollutants were led away to jail.

When Rachel returned from City Hall she was accosted by an official from the Florida Department of Revenue. "You'll have to pay sales tax on these cannon balls. They just passed a new law yesterday. Fill out these forms." It turned out she didn't have a peddler's permit, either.

As the state revenuer left, the IRS agent arrived. "This land transfer will trigger a capital gains tax." The general was becoming exasperated. He triggered the IRS agent with one round right through the middle of his forehead.

He then called for his finance director, who confirmed that indeed there would be a tax on the gain. The general also triggered his finance director.

Matters were growing desperate. As the sun went down the general ordered his troops to parade rest. They had been standing at attention all afternoon. Meanwhile the governor and his staff had all collapsed in the hot sun.

When the streetlights came on, you could see lawyers and bureaucrats and politicians lined up all around the plaza with forms and subpoenas and lawsuits in hand. They represented, among others, the following departments: Internal Revenue, Florida Revenue, Planning, Zoning, Concurrency Review, Future Land Use, Comprehensive Plans, Growth Management, Traffic, Sewage, Drainage, Environmental Regulation, Transportation, Building Permits and Asbestos Removal (the Spanish governor was wearing asbestos underwear).

Andy Jackson was by this time furious, but matters were much more easily solved in 1821. As Napoleon said, "You should take a [politician] out and hang him occasionally just to encourage the others."

The general tied the governor to the muzzle of a cannon and fired one round. His remains can still be seen spattered on the north wall of the Pensacola Cultural Center. Then, for the sake of economy, he lined up all the lawyers and bureaucrats and elected officials and fired one cannon ball through the whole mess.

The general refused to accept the Territory of Florida under such stringent conditions and marched back to Washington, D.C., with Rachel, to make a report. Rachel didn't like Pensacola anyway.

According to this scenario, we would still be speaking Spanish in Pensacola even to this very day. Pensacola would be known as Novabarcelona.

As a matter of fact, as Andy rode off into the sunset, he is reputed to have uttered those immortal words, "El toro poopoo."

Ole!

1492

He didn't discover us; we knew we were here.
American Indians on Columbus

Over five hundred years ago Columbus discovered America, or at least we thought he did. But suddenly there seem to be many questions and much controversy about who Columbus really was and if he really discovered America.

Historians are studying and restudying Columbus to try to dig up some dirt. Some actually now believe we were discovered by the Irish or the Welsh or the Norwegians or even the Japanese. Regarding the raging debates about Columbus, Mark Twain observed that "the researches of many commentators have already thrown much darkness on this subject, and it is probable that, if they continue, we shall soon know nothing at all about it."

Not to be outdone, we sent our reporter out to join the research, and he made a fantastic discovery: Columbus discovered Pensacola in 1491, sixty-eight years before DeLuna, who supposedly got here first.

Columbus sailed into Pensacola Bay on April 15, 1491, and tried to get into Bayou Chico, but the drawbridge was stuck in the down position. He couldn't land at Pensacola NAS (there was none in those days as the Indians had no Navy) so he sailed on to the Municipal Auditorium.

There awaiting him on shore, standing on a 55-gallon drum manufactured by Florida Drum Co., was Creek Indian Chief Pontiac Whibbs. "Well, here comes the neighborhood," sighed the chief as Columbus and his crew paddled ashore in rubber rafts. The chief then delivered for the first time his famous "eastern gate to the western state" welcome because that's where Columbus thought he was.

The parties smoked the peace pipe at the corner of Main and Tarragona and then bedded down for the night at the Waterfront Mission as the air conditioning was out at the T.T. Wentworth Museum and the San Carlos Hotel had been boarded up for years. The next day Columbus set out to discover the towns of Century and South Flomaton (they had not as yet merged in those days).

Unable to cross over from Davis Highway to Langley Ave. to avoid heavy traffic, Columbus next tried to find the I-110 extension north of I-10 but to no avail. Running out of travel funds, he marched back downtown and applied to the County Council of Braves for welfare. No county funds were available. In fact, Deputy Indian Chief Big Joe Flowers had already warned the county that Governor Super Chief Walking Dog Chiles in Tallahassee was threatening to capture the county.

Undaunted, Columbus sailed the Inland Coastal Waterway to Tallahassee seeking aid, but Walking Dog Chiles was also out of funds. There was no lottery in those days. Columbus returned, still empty-handed.

By this time the Indians were sick and tired of paying taxes to support Columbus. They scalped several of his crew and chased Columbus and his unscalped crew members back into the bay.

Homesick and weary of waiting for the passage of the additional one-cent sales tax, Columbus set sail back to Queen Isabella in Spain to seek wampum for future exploration.

Then came 1492, and you know the rest of the story: Columbus missed the U.S. mainland in 1492 because of his outdated Florida map which showed Northwest Florida cut off from the rest of Florida and parked out in the Gulf in the vicinity of Key West and Cuba.

To this day the Indians regret having dismissed Columbus so unceremoniously because they had hoped to retain the *Santa Maria* as an historical museum. The *Santa Maria* went to Corpus Christi instead, along with the U.S.S. Lexington.

Please amend your history books accordingly.

Foundations

Too much of a good thing is wonderful.
Mae West

The great humorist and actor Will Rogers was called upon to lecture on many subjects. Never bashful, on one occasion he was called upon to talk to the corset manufacturers. He was never at a loss for words.

Although this apparel is no longer widely used, Will's lecture will refresh your history of the good old days earlier this century:

"This speaking calls on a fellow to learn something about the articles that a self-respecting man has no business knowing about. So that's why I am going to get away. If a man is called on in a public banquet room, to tell what he knows about corsets, there is no telling what other ladies' wearing apparel he might be called on to discuss next. So get me back to the morals of Hollywood, before it's too late.

"I was, at that, mighty glad to appear at a dinner given by an essential industry. Just imagine, if you can, if the flesh of this country were allowed to wander around promiscuously! Why, there ain't no telling where it would wind up. There has got to be a gathering, or a get-together place for everything in this world, so, when our human bodies get beyond our control, why, we have to call on some mechanical force to help assemble them and bring back what might be called the semblance of a human frame.

"These corset builders, while they might not do a whole lot to help civilization, are a tremendous aid to the eyesight. They have got what you would call a Herculean Task, as they really have to improve on nature. The same problem confronts them that does the people that run the subways in New York City. They both have to get so many pounds of human flesh into a given radius. The subway does it by having strong men to push and shove until they can just close the door with only the last man's foot out. But the corset carpenters arrive at the same thing by a series of strings.

"Now the front lace model can be operated without a confederate. Judiciously holding your breath, with a conservative intake on the diaphragm, you arrange yourself inside this thing. Then you tie the strings to the door knob, and slowly back away. When your speedometer says you have arrived at exactly 36, why, you haul in your lines and tie off.

"We have also the side lace model that is made in case you are very fleshy and you need two accomplices to help you congregate yourself. You stand in the middle, and they pull from both sides. This acts something in the nature of a vise. This style has been known to operate so successfully that the victim's buttons have popped off their shoes.

"Of course, the fear of every fleshy lady is the broken corset string. I sat next to a catastrophe of this nature once. We didn't know it at first, the deluge seemed so gradual, till finally the gentleman on the other side of her, and myself, were gradually pushed off our chairs. To show you what a wonderful thing this corsetting is, that lady had come to the dinner before the broken string episode, in a small roadster, and she was delivered home in a bus.

"Imagine me being asked to talk at a corset dinner, anyway. Me, who has been six years with the Ziegfeld Follies, and not a corset in the show. Anyhow, men have gone down in history for shaping the destinies of nations, but I tell you this set of corset architects shape the destinies of women and that is a lot more important than some of the shaping that has been done on a lot of nations that I can name offhand. Another thing that makes me so strong for them, if it wasn't for the corset ads in magazines, men would never look at a magazine.

"Now, of course, not as many women wear corsets as used to, but what they have lost in women customers, they have made up with men."

How times have changed! Today in this modern world we just say, "Let it all hang out."

The Good Old Days

> One must wait until the evening to see how splended the day was.
> *Sophocles*

The Alger-Sullivan Historical Society in Century just held its annual meeting. These affairs always bring back fond memories of sawmill days and lumbering folks. Great stories, great history, great people!

One famous Century sawmill character was Probate Wright, a venerable sawmill hand who held a regular job in the sawmill plus doing extra yard work and odd jobs on the side. One day while doing some yard work, Probate announced he had to leave early. "Why?" we asked.

"'Cause I have to go down and draw my unemployment." We never did figure out how he got away with that deal.

Senator Robin Swift operated a sawmill in Atmore, Alabama. His wife told the tale about driving with the senator to Mobile one fine day. On the way he hit a cow, and the car ended up on top of the poor creature.

He couldn't back his car off the cow so he got out his jack to try and jack the car up over the cow. The jack wouldn't work, which infuriated the good senator who was a very fine southern gentleman with a very fine temper. He stomped on his hat and threw the jack as far as he could in a moment of unmitigated anger.

The jack landed in a tree. The senator had to climb the tree to retrieve his jack. His wife couldn't stop laughing, which added to the humiliation and indignity of the situation.

Another famous Alabama sawmiller was one Peter Vredenburg of Vredenburg, Alabama. Peter didn't appreciate some unannounced guests who showed up one day in his living room. Peter rounded up some of his fighting gamecocks and turned them loose in the living room. Those particular visitors have never been seen since.

In the old days at the mill in Century, the workers used to line up at the office pay window to be paid in cash. At the end of the line stood the Century loan sharks trying to collect; they charged 25¢ interest per $1 per week. That's 1300% interest. After being paid off, the mill hands tried to scatter on the run in all directions with the loan sharks in hot pursuit. After 1957 workers were paid by checks down at the mill. This change slowed up the local loan business considerably.

Historian Mabry Dozier, one of the company's management staff, tells the story about inspecting the company rental houses every year. Many workers resented this intrusion on their privacy, but it had to be done for safety and maintenance reasons.

One tenant complained bitterly about leaks in his roof. The poor man had buckets stationed all over the house to catch the drips. The man wanted a new roof, but the company said no, that was too expensive. So they just kept patching, and the roof just kept leaking.

The tenant was irate. Finally he said to Dozier, "You ask your boss if he has ever tried to sleep in a wet bed with a mad woman." The boss hadn't; the man got his new roof.

Another classic story involved the new owner, Leon Clancy, in 1957. On his first trip to Century Leon rolled into town in his big, long Cadillac, smoking a big, long cigar. He stopped at a small filling station-grocery store to buy some gas. Leon left the engine running and went inside. The clerk went out to pump the gas. After a while he came back into the store and said to Leon, "Mister, would you mind turning off your engine? You're gainin' on me!"

All of this history needs to be preserved. One thing for sure, "One page of history is worth a whole volume of theory."

ON LIFE

Marriage is the chief cause of divorce.
Groucho Marx

I'm not saying I'm heavily insured, but when I go, the insurance company goes.
Jack Benny

If I could drop dead right now, I'd be the happiest man alive.
Samuel Goldwyn

Make it idiot-proof and somebody will make a better idiot.
Anonymous

I wouldn't want to be a member of any club that would have me as a member,
Groucho Marx

You should always be careful when reading books about health. Otherwise,
you might die of a misprint.
Mark Twain

50 Years and Counting

You can't put your shoes on backwards and walk forward into the past.
Author Unknown

You haven't really lived until you have been to your 50th high school reunion. We have just returned from our spouse's 50th in Albany, Georgia. We attended as the "designated other," as the government would list us.

Paul Valeria once said, "The future is not what it used to be." How times do change!

Our old friend Mr. Pete Noonan used to say that there are three stages in life: youth, middle age, and "lookin' good."

Some of the 50-year graduates in Albany were "lookin' good." Some weren't. Some were looking better than others. Some weren't. Hearing aids were abundant. Hair was not. Some of the hairpieces were bald.

You couldn't identify anybody from the 1942 high school yearbook without a lot of help and guessing. Graduates painfully groped around through their bifocals trying to recognize old friends. Very few lived up to their pictures, predictions, and expectations printed in the 1942 yearbook.

Some old flames were rekindled, barely. Many of the graduates courted one another at one time or another and then went off and married somebody else.

Many of the 50-year greetings went something like this:

"Hi, you old goat. Good to see you again. You're lookin' good."

"Same to you, ole buddy. You're lookin' good. How's your sweet wife?"

"Oh, Helen's fine. Kids and grand-kids are all fine. How's your family doing these days?"

"Just great. Ginny and I have five grandchildren. They're all doing great. You remember Ginny?"

"Sure I do. She's a peach. Is she here?"

"No, not exactly. I'm here with my third wife. Meet Milly."

"Hi, Milly. Glad you caught up with this good old boy. He's the greatest."

"Thanks, pal. You're the greatest. Remember all those great times we had together?"

And so on and so on it went for about ten minutes, swapping all the great news about each other, their families, and their old friends. After they finally parted with big bear hugs, Milly's husband came over to us and gently asked, "Who in the hell was that guy?"

So much for fond remembrances and the good olde days. Refreshments were served, and the laughs and lies got bigger and bigger.

Finally there was the contest to pick Miss Albany High of 1942. Since we were fighting World War II in 1942, no beauty queen was chosen that year.

Because of present equal opportunity laws, 15 men and 17 women competed for queen. Five men made the list of finalists. None of the women made the cut. There must have been a mistake. We couldn't believe it. Apparently all the judges were blind.

More and more refreshments were served while the five male finalists stumbled around the floor in high heels for the final beauty parade in front of the judges who were by now chock full of refreshments. Facing these contestants was a terrible ordeal. The finalists were unbelievably ugly. We tried not to laugh because it was too pitiful to behold. One alumnus strangled while trying to choke back the laughter.

These same five burly lovelies played on the 1942 championship Albany High football team. According to Virginia Slims, "They've come a long way, baby."

Finally the refreshments ran out, Miss Albany High was crowned, the band played the alma mater and taps, and everyone went home. The judges were led away. Everybody was home in bed by 9 o'clock.

All in all it was great fun. They are already planning the next 50th in two separate locations, one a good deal warmer than the other.

Thirty-Nine and Holding

> I refuse to admit that I am more than 52,
> even if that does make my sons illegitimate.
> *Nancy Astor*

We are wrestling with the question of birthdays. We just suffered another one.

In some ways they're fine, and in some ways they're not. We prefer the strategy of those ladies who never get past thirty-nine years. As someone said, "Life is a terminal illness." There is no sense in rushing things.

Jack Benny always lied about his age. He used to talk about his life insurance coverage: "I won't say I'm heavily insured; but when I go, the insurance company goes."

There is that great story about Justice Oliver Wendell Holmes who at age eighty-nine caught sight of a very pretty girl. "Oh," he sighed, "just to be seventy again."

Later the justice was riding on a train when the conductor came by and asked to see his ticket. The justice fumbled around, embarrassed, but couldn't find it. "Don't worry, Mr. Justice," said the conductor helpfully, "I know who you are."

"That's not what I'm worried about," replied the justice, "I just want to know where I'm going."

One little boy asked his grandmother how old she was. "Ninety-two," she said. He was highly impressed but was forced to inquire, "Did you start at one?"

Two elderly gentlemen met on the street. One said to the other, "I'm sorry, but I just can't seem to remember your name." Replied the other, "How soon do you need to know?"

So you can easily see that older age has its problems. We decided to seek out the wisdom of great minds on the question of age to see if we could find some comforting thoughts:

Elephants and grandchildren never forget.
Anonymous

The secret of staying young is to live honestly,
eat slowly and not too much, and lie about your age.
Lucile Ball

All the things I like to do are either illegal,
immoral, or fattening.
Alexander Woollcott

You only die once but for such a long time.
Anonymous

Eternal rest sounds comforting in the pulpit. Well,
you try it once and see how heavy time will hang on your hands.
Mark Twain

So live that you wouldn't be ashamed to sell the family
parrot to the town gossip.
Will Rogers

Some comforting thoughts, some not.

Mark Twain summed it up best of all: "Let us endeavor so to live that when we come to die even the undertaker will be sorry."

Thirty-nine is much more comforting. Stay there as long as you can hold out.

JAWS IV

. . .if we didn't wear fur coats, those little animals would never have been born.
Barbi Benton, ex-Playboy Bunny

A shark recently attacked a surfer on the west coast of Florida. Man-eating sharks are something those of us along the gulf coast always keep in mind when immersing ourselves in the briny gulf liquid.

But who ever thought about a man-eating *squirrel?* That's correct: *squirrel.* We now have just such a monster in the neighborhood.

Last Sunday, a beautiful spring day, my spouse (or "designated other" in official government language) ventured out for an innocent afternoon stroll around our circle.

Little did she know what was lurking behind the bushes. She suddenly and without any warning felt a vicious bite on the back of her leg. Her first reaction was bear attack. Alone and unarmed she began hollering for help. In the meantime, the beast is chewing up and down on both legs and starting on her arm. She fell down.

Flat on her back on the ground, yelling for help, and flailing away with her arms, she suddenly and to her horror discovered she was being eaten alive by a squirrel, a large fox squirrel with large white teeth. Fortunately, before the attacker reached her neck and face, he broke off and retreated to the bushes; apparently he had eaten all he could digest for the moment.

Bruised, maimed, and bleeding, our attackee limped home with the aid of a nearby gardener. Rushed to the hospital by her horrified designated other, she spent five hours in the emergency room, hurting and still bleeding. She took rabies and tetanus shots in every conceivable part of her anatomy, including each wound. She was so full of holes, she was afraid she might leak. The shots continued for one month.

A former law enforcement officer, on hearing her cries, shouldered his trusty blunderbuss and marched off to stalk the killer and alert the villagers. An alarm went to the Pensacola police, who arrived on the scene armed and ready.

Lo and behold, smelling a chance for some new blood, super-squirrel charged forth once more and chased the city gendarme, who fought back gamely with his billyclub but missed subduing the mad monster. Nobody fired a shot for fear of shooting the policeman.

The animal shelter, environmentalists, health department—all involved bureaus have been alerted.

Zoo director Pat Quinn has been called in for consultation. He has organized an African jungle safari to comb the bush and hunt down this voracious beast. He doesn't believe it's a squirrel. According to the size of the teeth marks, Mr. Quinn estimated the creature may very well be a Bengal tiger. The victim said it looked and felt that big. Whatever it is, Mr. Quinn is anxious to put it in his zoo with the rest of the tigers. Traps are set. We await results.

Environmentalists are ecstatic that wild game is not extinct but is very much alive and well. The constabulary is cautious, having already been chased. Zookeeper Pat Quinn is hopeful for a successful capture. The health department is furious; rabies shots cost them $800.00

We don't know yet how the big game hunt will turn out. In the meantime, when feeding peanuts to your pet? squirrel, wear protective armor and have a hand grenade ready. A bulletproof vest is also suggested in case somebody is shooting at the squirrel. Do not venture out unarmed.

(Editor's Note: In response to anxious animal lovers and environmentalists who have inquired, the Health Department wishes to assure the public that the victim could not possibly have infected the squirrel.)

JAWS V - The End

That's the most unheard-of thing I ever heard of.
Senator Joseph McCarthy

By now you should have heard the terrible tale of the killer squirrel. There's more. Here is the end of the story.

After the first dastardly attack by super-squirrel on the squirrel lady (a person who shall remain nameless for fear of vengeance by the squirrel's immediate family), plans were carefully made and traps carefully laid to snare the beast. Alas! To no avail!

The trappers snared one raccoon, who was very upset and very angry at this indignity, and dozens of innocent gray squirrels, who were also mad as hell about the inconvenience, but no super-squirrel. He (or maybe she, we don't really know) cannily disappeared from sight, lying low until the storm could blow over. This squirrel is cagey.

Then, when the furor over his indiscretion had died down and the coast looked clear again, super-squirrel came back out of retirement and attacked again. This time he decided to munch on an unfortunate gardener tending to his chores in a backyard along the bayou. The squirrel prefers drumsticks. Fortunately the gardener was wearing long, heavy trousers, and super-squirrel could not get to his leg to draw any blood. The gardener finally beat him loose with his rake.

The gardener's trousers did not taste very good and did not satisfy the squirrel's gnawing hunger. After all, he hadn't tasted anybody for more than a week. This fact simply further infuriated our ravenous rodent. So he attacked the house. The house! The whole house! He wanted inside to wreak revenge on the owner.

Needless to say, the poor gardener had by this time retired from his post, permanently.

You can now imagine the surprise and consternation of the lady of the house to find a large red fox squirrel peering at her through the window and trying to break into her house, hungry and fully bent on getting another good meal.

Realizing that this was the same furry menace recently reported at large in the neighborhood, the lady quickly summoned reinforcements. The police and the FBI responded immediately, armed to the very teeth. They could not believe what they were seeing. In all the annals of law enforcement, there have been no reports of a squirrel attacking a house, let alone a gardener and a neighbor lady. Bears, maybe, but not squirrels.

The two policemen could find nothing in their instruction books to cover this situation; squirrels were not even mentioned. They called in for backup and further instructions. Headquarters checked the files and police manuals and discovered a law against shooting squirrels, which was not of very much help in light of the present emergency. There were no references to attack-squirrels, only pets.

The FBI phoned in to Waco, Texas, for advice. Waco had nothing covering squirrels but recommended an M-60 tank and tear gas. Attorney General Janet Reno approved; there just wasn't time to notify the president.

The police gave up the chase, retired to the front yard, hid in a clump of bushes, blindfolded each other, and plugged up their ears.

A shot rang out! Bull's-eye! Right through the heart. Mortally wounded, super-squirrel clutched his heart, staggered, waved goodbye and entered eternal rest right then and there. Certainly this was far more merciful than hanging or electrocution after a long trial. The shooter shall forever be unknown to all except God. None of the police saw or heard anything. It was "see nothing, hear nothing, know nothing."

The police sacked up the victim in a plastic evidence bag and sadly retired from the scene with the corpse. The reign of terror has ended. Mothers and babies can now once more confidently venture out, secure in the knowledge that they are safe from further attacks by this roving predator. The siege is over!

One For The Road

> We are in a pretty mess; can get nothing to eat,
> and no wine to drink, except whiskey.
> *Sir Boyle Roche*
> *(Commenting on a visit to Ireland)*

Prohibition didn't work as we still have drinking and drunks. Drunks are a menace to themselves and others and need to be controlled. They can get themselves into the darnedest messes.

In Louisville a paraplegic driving his battery-powered wheelchair home from a bar was arrested for driving-under-the-influence (DUI). The judge is not sure what to do with this case. Another gentlemen was arrested while driving and drinking on his power mower. There are reports of many horseback riding accidents caused by drinking (by the rider, not the horse). Is this DUI?

These cases would be funny if not so serious, but there are a lot of good jokes about drunks.

There was the drunk leaning on a telephone pole and feeling his way around it. Finally, in desperation, he muttered, "It's no use. I can't get out. I'm walled in!"

Then there were the two drunks walking along a railroad track. One complained, "This is the longest stairs I've ever been on." Said the other, "I don't mind that, but the handrails are too low!"

Another drunk staggered up out of the subway and exclaimed, "Jeese, you ought to see the big set of electric trains that guy has down there."

A friendly drunk put a nickel in a parking meter and gasped, "Good God, I've gained 100 pounds."

Another drunk started to get into his car and drive off when confronted by a policeman. "Surely," said the cop, "you're not going to drive in that condition." "Well, Officer," said the man, "I'm sorry, but I'm certainly in no condition to walk."

There was the drunk driver who was stopped for driving the wrong way on a one-way street. "Didn't you see the arrows?" asked the policeman. "No, sir," said the driver, "I didn't even see the Indians."

A heavy beer drinker fell out of the fourth floor of a building and landed in the street. A crowd rushed up and asked him what happened. "I don't know," said the man, "I just got here myself."

Then there were two drunks fishing from a boat at night. One looked down and observed the reflection of the moon on the water.

"What's that down there?" he asked.

"That's the moon," answered his partner.

"Well, if that's the moon down there, what are we doing up here?"

Another drunk was standing up at a bar next to a grizzly bear. When he put his arm around the bear, the bear threw him out into the street. Complained the drinker, "Give some women a fur coat and they think they own the world."

The most pitiful case was the wife who didn't know her husband was a drunk until one night when he came home sober.

Well, all of this would be funny if drunks weren't so dangerous. We need to keep them off the streets and off of horses and in the comic books.

Cheers!

Pull the Plug!

> It wasn't raining when Noah built the ark.
> *Anonymous*

We have drainage problems which can turn into floods.

Our county is not as flat as South Florida, where you can stand on a stepladder and see 100 miles. But we do have a lot of flat places where the water doesn't have any place to go except up, out, into a tidal wave, and then into your shoes, house, outhouse, car, or barn.

We took the ferry during a heavy rainstorm to make a report. Jonah J. (for Jehosophat) Jones served as our guide in a rubber life raft. We all wore life jackets and carried radios, flares, snorkels, and other deep water emergency gear. We stood in our hipboots knee-deep in the flood waters on Jonah's front porch. His furniture was floating quietly out the back door.

"We hope this is as high as it gets," we said as the water lapped at our belt buckles.

"We ain't even started," said Jonah sadly. "Man the lifeboat."

Sure enough, a wall of water appeared from around the bend in the road, heading straight for us. Jonah pulled up the anchor attached to his house, and the house floated quietly away downstream following his furniture.

We moored our raft safely to the top of a pine tree and watched the small ocean as it spread out at high tide.

"How do you get your house back?" we asked Jonah.

"It's tied to this here pine tree with a line wrapped around my fishing reel. When this here water starts going down, we just reel it back in and drop anchor again in this here same spot. Nothin' to it, happens once or twict a week." He looked disgusted.

As he spoke, a two-story house and several cows floated past. Next in the parade came a complete moonshine still with the operators paddling madly. Right behind them paddling furiously in hot pursuit came a sheriff's posse firing cannon shots across the bow of the still. We never did find out who won the race.

As the floodwaters receded, we disembarked and waded to a telephone to file our report while treading water in the phone booth. Many of the neighbors were swimming back home and sorting out their houses, which bobbed around like corks on the waves.

We asked Jonah what the solution is to our continued drainage problems.

"Dig a lot more ditches, build some hills, or else just pull the plug," he said damply.

"If we can find the plug, we'll pull it," we said. In the meantime, all you flatlanders keep your boots and lifeboats handy.

The Higher They Go

> Agrogooberacrophobia - Fear of peanut farmers in high places.
> *Unknown*

There has been a rash of news lately about people falling out of the sky.

President George Bush jumped out of an airplane with a parachute just for old times' sake. He was a WWII pilot who jumped into the ocean on one occasion.

Then our famous Brownsville evangelist fell off his house and busted himself into several pieces. House building can be injurious to your health if you don't hang on tight.

We have been watching workers reroof a neighbor's house. These young daredevils walk about perilously close to the edge of a 2-story roof with no safety harness or nets in case of a plunge. They are too nonchalant for us. We just close our eyes and pray. We get acrophobia just watching them teeter on the brink.

Sometimes people are pushed off high places; sometimes they jump. Late news told of a lady looking out her second story window and seeing a pit bulldog chewing on her very own dear pet poodle. Panic set in.

With no time to spare she jumped out the window to save her darling dog. She broke her ankle, but undaunted she bit the pit bull. *BIT him*! This heroic sacrifice so startled the bulldog that he let go and took off for dear life before she could chew on him any more. The rescued poodle is still alive and well. We don't know about the pit bull.

We contacted the Humane Society to be sure the lady received a lifesaving medal. We were shocked by the response.

"We're not handing out any medals on this case. We've had nothing but grief ever since it happened."

"How can that be?" we asked.

"The animal rights people got all over us. They're mad because she attacked the bulldog, and they're afraid it may get rabies. They want the lady caught and tested to make sure."

"Good Lord," we said.

There are other stories about great leaps and falls. One heavy drinker full of beer fell out of a three story building and landed on the pavement below. A crowd rushed up.

"What happened?" asked one bystander.

"I don't know," said the victim, "I just got here myself."

Mark Twain tells the story of helping rescue a man from the third story of a burning building. Twain arrived on the scene and, seeing the man's dire predicament, threw him up a rope. Twain said he pulled the man down to safety without any trouble.

Finally, there is the tale of the pilot who bailed out of his burning plane and then discovered that his parachute wouldn't open. On the way down he passed another man who was shooting up past him at a rapid rate. The pilot called out to the other man, "Say, fella, do you know anything about parachutes?"

"No, sir, I'm sorry," came the response. "Do you know anything about gas heaters?"

What goes up must come down harder!

Into The Air

> If God had intended for us to fly,
> he would have made it easier to get to the airport.
>
> *Unknown*

What goes up has to come down somehow.

There seems to have been a rash of airplane crashes recently, but flying still remains much safer than driving your own car. In fact, even with all the military crashes during the past month, the services' record is still down one-third this year.

So take heart, think about the bright side, and enjoy some flying stories.

Four men were flying together on an airplane: the president, a professor, a clergyman, and a young hippie. The plane developed trouble and the passengers had to bail out, but there were only three parachutes. What to do?

"I have some vital national duties to attend to so I'd better jump," said the president. So they gave him a chute.

"I'm the smartest man on board, and I have important knowledge that we can't afford to lose," said the professor. So they gave him a chute.

That left only one parachute.

"You're young and have your whole life ahead of you," said the clergyman to the hippie. "You go ahead and take the last chute."

"That won't be necessary, sir," said the young man, "that smart guy jumped out with my knapsack." That ended that emergency.

Four other men were on a deer hunt and killed six deer. When they were ready to go home in their small plane, they wanted to put all six deer on board. "The plane is too small. It won't carry that load," warned the pilot.

"Well, we did it last year," protested one of the hunters. So they loaded up all the deer and took off. The plane soon crashed.

They all survived and climbed out of the wreck. "Where are we?" asked the pilot."

Said another one of the hunters, "We're just one mile from where we crashed last year."

Everybody kids about the airlines: food, baggage, service, costs, schedules, late. As one wag put it, "Lindbergh was the first to fly the Atlantic alone and the last to arrive with his baggage."

On board one scheduled airline flight, the stewardess was taking orders for drinks. When she asked one particularly religious gentleman what he would like, he was highly insulted. "Young lady, I'll have you know I would rather commit adultery than take a drink of whiskey."

An inebriated gentleman sitting just behind him stood up and said to the stewardess, "Young lady, cancel my order for a drink. I didn't know we had a choice." To each his own.

Passengers arriving in Pensacola years ago swear that on the approach to the airport, the pilot came on the radio and announced, "Ladies and gentlemen, we will soon be landing in Pensacola, Florida, which is on Central Standard Time. Please set your watches back twenty-five years."

Flying isn't really that dangerous. It can also be a lot of fun, and sometimes get you there on time, with or without a drink.

Hog Heaven

> Odd things, animals. All dogs look up to you. All cats look down at you.
> Only a pig looks at you as an equal.
> *Winston Churchill*

Farmers usually stick together, but there's an agricultural civil war brewing in Iowa.

The pig farmers and the other farmers are fighting. In one case a large pig farm is raising 6000 pigs all jammed in like sardines under one roof. That's a lot of bacon. Unfortunately, it's also a lot of pig manure.

The pig farmer collects his manure daily and spreads it all over his farm to raise his other crops. Whew! According to his neighbors, the stench is terrible. One protesting neighbor is two miles away, and he can't stand it. He's ready to move or kill the pig farmer.

Naturally the situation is tense and may lead to armed conflict. In an attempt to mediate and avert open warfare, we went to see dairy farmer Angus Kowpize who owns a farm close to the pigs. Angus was holding his nose; his wife had a clothes pin on hers.

"Can we make peace among you farmers?" we asked.

"Are you kidding? Can't you smell it? They're ruining the whole neighborhood. My house smells like a pigpen."

"What can we do to help?"

"Nuthin' I know of. We've tried everything. My whole family is wearing clothes pins. My cows are mad as hell; we had to put gas masks on 'em."

"Have you tried to visit the pig farm to make peace?"

"You must be joking. You can't get within a mile of that stinking hole. It's worse than a gas attack. Go over there and smell for yourself." He ran off shouting "Gas" to anyone who would listen.

So we went to visit Herman "Piggy" Hogg whose farm raised all the stink. We'll have to admit the aroma was pretty potent. Hogg himself was on oxygen and handed us an oxygen mask.

"We've come for a truce. Is there some way we can cut down on this manure?" we asked.

This agitated Hogg. "You think I enjoy the stink? But what do you expect 6000 pigs to do? They can't hold it; they have to go someplace. Nature calls all 6000 of 'em. But I love my pigs. Aren't they cute?"

"How do you stand it?"

"We don't inhale."

"How do the pigs stand it?"

"Well, you know pigs."

"Is there any way they could be potty trained?"

"What? 6000 pigs on potties? You must be nuts. You city slickers need to get out here and enjoy nature."

He obviously enjoyed his work. "Here piggy, piggy, piggy," he called and then sprayed them with deodorant and Chanel No. 5. It did not help.

We whiffed all we could stand and left in search of fresh air. As we drove off we noticed some of the pigs were holding their noses.

We now understand the problem, but we don't see any answers. We have to have bacon to go with our eggs.

"Jump": The last word in airplanes

Do unto others before they do unto you.
Anonymous

Modern airplanes are marvels of power, electronic gadgets, and skill. If you have ever looked into the cockpit of a jetliner, you have seen row upon row of dials, gauges, switches, handles, and levers.

Things were different in the good old days when Lindbergh flew the airmail. Planes were put together with canvas, bailing wire, and chewing gum, not much metal and not many instruments. Mainly just a "go and stop" handle, a control stick, and a compass.

Things had improved somewhat by early World War II. There were some metal training planes with a few more instruments, gauges, buttons, and handles. We had an artificial horizon which told us whether we were upside down.

There was also a radio compass for direction finding. You tuned in to a radio station, and a needle pointed the way. But during bad weather the darned thing homed in on thunderstorms. Not very comforting.

There was no radar or distance finding or automatic direction finding or other modern equipment, particularly in training planes. You had to figure out where you were on a map. But those were the exciting days of airplane driving, more fun, more seat-of-your-pants flying, fewer regulations, more hell-raising.

Thousands upon thousands of pilots trained during WW II. The poor, harried instructors had to risk their lives with all kinds of characters, wildmen, and lunatics.

Some primary training planes didn't have radios, just a hose acting as a speaking tube between the student and instructor. A favorite stunt of instructors if you made a mistake was to stick the tube out into the slipstream. This maneuver almost blew the head off of the poor student. You didn't make the same mistake twice if you could help it.

There is the classic tale, true or not, about the instructor pilot who, just before each student's first solo flight, always played a little trick on the poor unsuspecting cadet.

As he and his student were flying along happily and peacefully together, the instructor unhooked his control stick and threw it overboard, much to the student's consternation. But the instructor figured this was a good way to instill confidence in the fledgling pilot. He was now on his own.

One particularly playful student found out this was going to happen to him so he secretly took along an extra control stick. He was ready! When the instructor tossed his stick away, the student threw his extra stick overboard. The instructor bailed out.

Another playful student in advanced training used to hover over the field. When one of his unsuspecting fellow aviators was just about to land, our villain would grab his microphone and radio to his classmate, "Airplane on fire, pull up and go around."

The poor victim trying to land in peril was in a terrible predicament. Should he crash-land on fire or should he pull up and burn up in the air? The culprit was never caught. Nobody ever actually burned up.

Another flying cadet got lost one day on a practice cross-country. When he was about to run out of gas, he made an emergency landing in a farmer's field. He crawled out of his airplane and approached the farmer who was plowing his field.

"Where can I find a telephone?" he asked the farmer.

"Dunno," said the farmer.

"Well, which way is the closest airport?"

"Dunno."

"Well, how do I get to town?"

"Dunno."

Finally the exasperated student pilot couldn't stand it any longer. "Mister, you don't know a damned thing, do you!"

"Well," said the farmer, "I'll tell you, sonny, I know one damned thing for sure. I ain't lost!"

The good old days have gone forever. Thank God!

Funnier than Fiction

> Comedy is tragedy that happens to other people.
> *Angela Carter*

The truth really is better than fiction, and funnier.

You've heard the story of Wrongway Corrigan. Years ago Wrongway planned to fly solo to Europe. The officials turned down his application so he announced he would return to Los Angeles and climbed into his airplane. The next thing we knew he was sighted over Ireland.

Wrongway claimed he had set his compass backwards. He didn't explain why he flew over so much water between New York and Los Angeles. A likely story!

Recently Florida Governor Lawton Chiles flew to Panama City to make an appearance for the Democratic state senate candidate campaigning there.

The governor's plane landed gracefully and stopped on the tarmac. His aides stepped out onto the pavement and looked around. They were mystified. Where was the VIP welcoming committee? Where was the crowd? Where was the rally?

Finally the governor himself climbed down, stretched, and looked about. He, too, was mystified. His speech was ready and rehearsed. Where was the campaign?

"Where the hell is everybody?" he inquired. Everybody shrugged. Nobody knew. Embarrassed silence. Then the governor himself happened to look up at the sign hanging on the front of the hangar: "Welcome to Pensacola, Florida."

"Good God," said the governor. "We're in the wrong town." He, his crew, and his aides remounted their aircraft and flew back to Panama City which they had passed over 30 minutes earlier. Unidentified sources indicate the governor gave the crew 40 lashes and is now advertising for a new navigator.

A short time ago a high school football team asked the Army parachute team to bail out over the school, land on their field, and deliver the game ball in an exciting opening ceremony.

All was in order as planned. The parachute team jumped on schedule and made perfect landings right in the middle of the well-lighted field. Game ball in hand, the team leader approached the officials to hand over the ball.

"Who are you?" asked the head official.

"We're the Army. We have your game ball," the team leader announced proudly. "You can't play without a ball."

"We've already got a ball."

"Well where in hell are we? What game is this?" The official told him.

"Oh my God!" said the team leader. "We landed in the wrong stadium." The team finally delivered the ball to the right stadium in a taxicab.

In still another city a nice lady drove along peacefully when she heard a shot. She felt something sticky on the back of her head. She reached back and felt a wet glob of gooey stuff oozing out of the back of her head.

"Good Lord, I've been shot. My brains are coming out," she realized. The police roared to the scene to investigate. She still held on to her brains for dear life.

In the back seat of the car the police discovered a can of biscuit dough which had exploded. She was holding onto a soggy biscuit which had stuck to the back of her head. There was a large sigh of relief all around. She picked up the rest of her biscuits and drove home without further injury.

There's nothing like being in the wrong place at the right time.

ON THE MILITARY

You should take a general out occasionally and hang him just to
encourage the others.
Napoleon

Git there firstest with the mostest.
Nathan Bedford Forrest, Confederate General

The more we sweat in peace, the less we bleed in war.
General Omar Bradley

Peace is the dream of the wise; war is the history of man.
Unknown

A Gathering of Eagles

The past is the only dead thing that smells sweet.
Cyril Connolly

We recently attended a reunion of our flying class of 1945, the '45 Eagles. A report on this drooping, graying, balding, forgetful, bespectacled, slipping, sagging, bulging, paunchy, dragging, arthritic assemblage of tired air jockeys is in order.

The meeting consisted of the usual reunion fare: drinking, hangar flying, tall tales, lies, some minor truths, and drinking. Each of us was supposed to get up and recite a *short* personal history. They couldn't shut us up.

Once at the podium we gazed out over an assorted sea of old and unfamiliar faces: full generals, lieutenant generals, major generals, brigadier generals, colonels, big industrial executives, distinguished public personalities, authors, artists, and the rest of us. It seemed appropriate to recall the sacrifices that the rest of us made so that the others could succeed. We made them look good! Here's how some of the rest of us did it, classmate by classmate, eagle by eagle:

We were the last to solo in primary so that Jack Broughton wouldn't look bad (he had a bad instructor). Little did we realize then that our heroic sacrifice would produce the future leader of the Air Force Thunderbirds.

He was also the hot jet jockey who disobeyed President Lyndon Johnson and pursued the North Koreans across the Yalu River against the president's orders. The president was fighting the war from his bedroom. The president then quit, but Jack didn't. Jack is now an author and hero. We did it!

Next we managed to groundloop in basic flying training. This is a horizontal loop while still on the ground. It is not a distinguished military flying maneuver.

In advanced training another eagle ran his airplane into the back of our airplane and chewed off the tail and the whole rear end right up through the rear cockpit. Fortunately we were in the front cockpit, which was all that was left. The chewer flunked his eye exam, washed out, and was sent to bomb disposal school where he accidentally blew himself up.

The last personal recollection involves a fellow eagle by the name of William T. Bess. We were assigned to fly B-17's with Mr. Bess. We personally, with him as co-pilot, taxied our B-17 into a telephone pole. This maneuver so humiliated Mr. Bess that he resigned in disgust and became executive vice-president of the gigantic Union Camp Corporation. We did it again!

By now you can begin to fully understand the significance and extent of the major self-sacrifices we're talking about here. Read on; there are more. Other eagles willingly cooperated in this self-immolation program.

Frank Lee flew the first Constellation aircraft equipped as a radar scanner. A gigantic welcoming ceremony waited at the San Francisco Airport for Lt. Lee to arrive on the maiden voyage. Bands played; high brass arrived; speeches were ready. The fog came in.

Time went by. And went by. Fog but no Frank. Suddenly through the mist Lt. Lee and his crew came rowing ashore in rubber rafts. Our hero had inadvertently landed in San Francisco Bay. So much for radar. Scratch one Constellation. Frank later sold hearing aids.

Red Dog Brenneman guarded a latrine door while Rocko Brett put in his contact lenses so he could see the eye chart. Rocko passed and ended up a Lt. General. He later flew with his aircraft's glass canopy ground to his prescription. Red Dog has disappeared from sight.

As we all struggled to learn to land, an anonymous eagle, probably Samuel Adams (no relation to the founding fathers), hovered high over the field preying on struggling young aviators. Just as some poor eagle was fluttering in for a landing, Sam would pounce, calling out on his radio, "Aircraft on fire, pull up and go around." This is traumatic to a fledgling. Should he land and crash or go around and burn up in the air?

One dark, stormy flying night a lot of us got lost on a cross-country trip. One poor lost eagle radioed, "Mayday! Help! I'm lost! I need assistance."

The tower answered, "Aircraft reporting lost, please identify yourself."

After a long pause came the reply, "Sir, I'm lost, but I'm not that lost." When the tower asked all planes on the final approach to turn on their landing lights, lights came on for 50 miles around. Somehow we all got back safely.

Pooky Brewer tried to land with his wheels up and locked. Humiliating but not fatal. This is an unauthorized flying maneuver, not required to graduate. Pooky was reduced to buck-private for this indiscretion.

On the night of our very first reunion in Germany, Scabber Avery ended up swinging from the chandelier in the club. A colonel reminded him of that fact the next morning. "Just thought I'd mention it," said the colonel. Scabber pleaded guilty and rehung the chandelier.

Bull Bartron, a supposedly sturdy classmate, played fullback on his airbase football team. In his final game, Bull broke loose at his own ten-yard line and headed untouched for the goal line 90 yards away. At the 50, in the clear all by himself, he slowed down. At the 25 he slowed visibly. At the 10 he collapsed from exhaustion and was borne off the field in ignominy by stretcher bearers. He finally regained consciousness and later recovered. No touchdown. That ended his football career, but henceforth he was better known as Rubberlegs Bartron.

Hairbreath Horowitz got lost one dark night and tried to land on a lighted bridge that looked just like a runway. Frantic natives waved him off.

Hairbreath then ran out of gas and landed on a city street. When his right wing came off, he turned left. When the left wing came off, he turned right and through the kitchen door of a family just celebrating the return of their soldier son from Japan. The family had fortunately just departed from the kitchen when Hairbreath drove in, unannounced.

Imagine their surprise. Our hero climbed out of what was left of his airplane and asked for directions to the nearest airport. The poor son was shell shocked, not from Japan but from Hairbreath's sudden entrance. He recovered with the help of doctors. Hairbreath rebounded to become a successful fiction writer. No wonder! This was a great start.

The last-but-far-from-least report is one sent to us later by our former roommate, Pasha Basham. Pasha tells of his flying assault on a tow-target during gunnery practice. During his attack Pasha was forced to land to go to the bathroom.

While in the bathroom two other intrepid airman riddled Pasha's target by mistake. Pasha got full credit for this spectacular gunnery score while seated on the commode. He received commendation and promotion for his deadly accuracy in the air.

(Editor's Note: These stories are approximately true. Names have not been changed to protect the guilty or to avoid prosecution.)

Bombs Aweigh

> You always write it's bombing, bombing, bombing.
> It's not bombing, it's air support.
> *Colonel David Opfer, USAF*

The Navy Tailhook affair in Las Vegas just won't go away. People still wonder what really happened. We sent our ace investigative reporter out to get the real facts. We will end this matter once and for all.

First we interviewed Lt. Tiger Pultenjees, a tailhook fighter pilot, at his base. We found him hangar-flying, doing some loops and slow rolls with his arms and hands just to keep in practice while being investigated.

"Were you there?" we asked.

"I can't talk. We're all under investigation. I'm saying nuthin.'"

"Can you just give us a hint?" we pleaded.

"Not a clue. I'm sayin' nuthin.' Go talk to the damned Air Force bomber pilots."

We didn't quite understand what Air Force bomber pilots had to do with anything, but we did follow up on this lead. We went to see Colonel Ariel Droppe, president of the National Pickle Droppers Association, a nationwide federation of old bomber pilots.

"What's the scoop on the tailhookers?" we asked Col. Droppe.

"Shhh! It's so classified I don't even want to talk about it," he whispered.

"Well, it's all going to come out anyway, so you might as well let it all out now," we warned.

He motioned us into a nearby phone booth where he whispered that he could talk anonymously if we wouldn't tell anybody. So here is the story, anonymously:

Fighter pilots have always made fun of bomber pilots. Fighter pilots have all the fun, get all the glory, and get all the girls. These are the macho guys who eat their Wheaties. Top-gun warriors. This glamorous reputation infuriates bomber pilots who plod through the skies driving the equivalent of flying moving vans. Not too thrilling!

So it seems that at the annual meeting of the Pickle Droppers Association, the bomber pilots came up with a diabolical plot to discredit fighter pilots. They aimed to get even.

First they smuggled a bunch of bomber pilots into the Tailhook convention in duffel bags. These mean spies kidnapped some Navy ladies and then started a riot on the third floor of the Hilton Hotel. The brave fighter pilots were alerted and innocently came to the rescue in an attempt to quell the riot and protect the ladies present. At the height of the battle, the treacherous bomber pilots sneaked out and left the noble and courageous fighter pilots holding the duffel bags and all the blame. So now you know the whole truth.

The Tailhookers vow revenge. Security at their next convention (but not at the Las Vegas Hilton) has been doubled to prevent a recurrence of this perfidy. They also plan to shoot down the bomber pilots. A group of small fighter pilots is now in training to infiltrate the next Pickle Droppers convention and start trouble. It's not over yet!

The Media Have Landed

> Do something, dammit, even if it's wrong.
> *General George Patton*

The press has been roundly criticized for its recent role in the U.S. military landings in Somalia. We wish to come to their defense.

According to the press version of the story, a SEAL-Marine landing team sneaking ashore in Somalia at night on a secret mission got lost in the dark. In desperation the team leader shouted out, "Has anybody got a light?"

Sure enough, the members of the press, who had already landed, were standing ready with dozens of flashlights. They turned them on all at once. What the fearless press was doing was drawing all the enemy fire away from our troops and onto themselves. They saved our SEALS. They are heroes.

The Marines were then able to find the airport, only to discover that it had already been captured by the press, who were interviewing the natives and handing out Girl Scout cookies. They applauded the Marines when they arrived. Our commandos were embarrassed by all the attention but reported in to headquarters that the airport had been captured. They didn't say by whom. Then the press further embarrassed the Marines by interviewing them on national prime-time TV.

There is ample precedent for such military operations by the media. They have been instrumental in winning battles throughout history.

At Waterloo, the press was interviewing Napoleon when he should have been out leading his troops. Then the English attacked. Napoleon, who was a real publicity hound, was so busy telling the reporter how great he was that he forgot to give his troops the order to shoot back. The enemy troops overran Napoleon's troops, and the rest is history.

When Napoleon discovered what had happened, he reached into his vest for a pistol to shoot the reporter. His hand got stuck, and that's why you always see pictures of Napoleon with his hand tucked in his shirt.

The media rendered great assistance to General Douglas MacArthur when he waded ashore in the Philippines during World War II. Reporters were already on the beach with dry britches for the general. The press had stormed ashore earlier and established a beachhead to insure a safe landing for the general. They then rehearsed the wade-in with the general for several hours until everybody got it right. They ran out of dry britches, but the story and pictures made all the front pages.

The press was equally effective in Iraq. Most of the Iraqis surrendered to the press during Desert Storm so they wouldn't bother General "Stormin' Norman" Schwarzkopf.

During the heat of the battle a reporter tried to interview the general while he was talking to General Colin Powell, his boss in Washington, D.C. General Schwarzkopf finally became so exasperated with the reporter's interference that he finally hollered to an aide, "Will somebody please shoot this little bastard!"

General Powell thought Schwarzkopf was going to kill Saddam Hussein and stopped the war because he had no United Nations authority to shoot Saddam. The press were already on their way to capture Saddam and now blame Powell and Schwarzkopf for calling off the war and interfering with their combat mission.

We must admit the media haven't always been so successful in combat. At Little Big Horn the press failed to show up for a scheduled press conference with Big Chief Sitting Bull, so Sitting Bull got mad, rode off over the hill in a rage, and took it out on poor General Custer. Custer was never the same after that.

The press does make lots of mistakes, but there have been other incidents where these mistakes turned out for the best. You will remember the famous naval incident involving Admiral Farragut. The Admiral spotted a torpedo coming straight toward him and hollered out, "Damn, torpedoes, full speed astern."

A reporter, who had spotted the torpedo coming before the admiral saw it, jumped overboard in a panic and swam to shore. In his haste he forgot exactly what the admiral had said. When he got to shore, cold, wet, and scared, he reported that Farragut's exact words were, "Damn the torpedoes, full speed ahead." Fortunately this misquotation gave us a great naval hero.

So you can see the important impact that the press has had on military history. Frankly we think the press has gotten a very bum rap. They should get the full credit they deserve.

The Hollow Forces

> God is on the side of the heavier battalions.
>
> *Voltaire*

Congress grows concerned about the readiness of our armed forces, and President Clinton has proposed to give the services an extra 25 billion dollars over five years. Republicans are howling for more. 25 billion will go through the Pentagon like green apples.

Can things really be this bad? General Bradley once warned that "the more we sweat in peace, the less we bleed in war," but we're already bleeding. In the 1980s Reagan rebuilt the armed forces to scare Russia into *perestroika* and *glasnost* and Yeltsin. Now that we don't have any big boys left to fight, we seem to be disarming again, according to critics. We are reminded of 1941 when the U.S. was caught by surprise, unprepared to fight World War II.

We decided to visit our troops in the field to find out what's going on. First we went to an Air Force hangar where we encountered Lt. Col. Hap Arnold Gonzalez, bomber squadron commander. He was running in place, flapping his arms up and down.

"What are you doing?" we asked.

"I'm getting in my flying time," said the colonel. "I'm just warming up the engines for takeoff." He took off running around the hangar joined by other squadron bomber pilots, all flapping their arms.

When the Colonel hollered "Bombs away," they dropped cans of tomato soup on tanks, factories, airfields, bridges, bomb dumps, and civilians painted on the hangar floor. They landed in formation. They all looked combat ready, but we couldn't tell for sure.

Next we went to see the Navy. The fleet was on maneuvers. We counted a dozen kayaks, 27 canoes, and assorted rowboats. Admiral David Farragut Jones was in command, standing in the bow of the lead rowboat.

"What's all this about?" we wanted to know.

"Damn the torpedoes, the Pentagon, and Congress," shouted the Admiral through his bull horn. "We have just begun to fight; but we don't have any torpedoes, and all my ships are in mothballs." The fleet was tossing pineapple depth charges overboard in an attempt to knock out imaginary enemy submarines.

The Admiral was in no mood to be bothered, so we saluted the bridge and went off to visit an Army post.

Regimental Commander Colonel Ulysses S. Grant greeted us sadly. He was riding in a golf cart labeled "TANK." The Colonel was carrying his bow and arrows and looked thoroughly disgusted.

"What on earth is going on?" we wondered out loud.

"Our tanks are grounded. The Pentagon is out of gas. I have to charge up the battery on my tank at home."

Up drove Sgt. Stonewall Jackson on a scooter labeled "halftrack." We rode with Sgt. Jackson out into the area where the troops trained.

They were conducting target practice using broom handles. The soldiers shouted Bang! Bang! Bang! and then waited anxiously for their scores.

Corporal Robert E. Lee's troops drilled with their broom handles at right-shoulder-arms. They then marched out to the range where they fired stove pipes labeled "mortars." They loaded the mortars with grapefruit and then lit firecrackers to get the effect of real combat.

"How accurate are these stove pipes?" we wanted to know, so we asked Private Armistead Pickett.

"Pretty damned good," he bragged. "Last week we knocked out three golf carts—er—uh—tanks."

Milking the System

> A cow may be drained dry; and if the Chancellors of the Exchequer persist in
> meeting every deficiency that occurs by taxing... industries, they will inevitably
> kill the cow that lays the golden milk.
> *Sir Frederick Milner*
> *British Parliament, 1900*

The U.S. Naval Academy owns and operates a dairy farm! The government bought this dairy years ago to insure that the midshipmen drink good, safe, wholesome milk.

The Navy has been operating this dairy at a cost of $1.5 million annually. Now the government has discovered that it can furnish good milk to the middies by buying it on the outside market for one-third of that amount or only $500,000 per year.

So the government has decided to sell the dairy and save $1 million per year. In addition, the Navy can now sell the dairy farm for a very handsome sum, in fact enough to buy milk for the middies for the next 100 years. The property is very valuable. This all sounds like a brilliant idea.

Taxpayers nationwide will rejoice at this government economy. Everybody, that is, except the folks in the community where the dairy is located. They are up in arms over the idea. We interviewed the mayor to find out why. These are his honest-to-God answers as published in the press and seen on TV:

"Why on earth would you be against this deal?" we asked his honor.

"I'll tell you why. Because this is the only dairy left in this part of the world. We're attached to it. It's historic. You don't believe in selling our history, do you?"

"That's pretty expensive history," we answered. "Surely the Smithsonian or some other museum can stuff some cows and preserve dairy history for a lot less than $1.5 million a year," we suggested.

"Well, you'll have to ask the cows about that," he said. "How would you like to be stuffed? But there are other good reasons. We don't want that property developed into a lot of apartments, shopping malls, bars, infrastructure, slums, and so on. We want to save the pristine beauty of the wide-open green spaces. This dairy environment is sacred to us."

"What about the dollars we could save?" we wondered out loud.

"It's just a measly million dollars a year plus 50 million for the land. That's a drop in the bucket compared to the trillions the government spends," his honor proclaimed, waving his arms emphatically. He was highly agitated.

We could see we weren't getting very far with the mayor. "What about all those cowpies out there? You could get rid of that mess," we suggested.

With that the mayor threw down his hat, stomped on it, and stormed off to city hall. We were just trying to be helpful.

This is one time the government is dead right. Even the cows agree.

Down to the Sea in Ships

I like to buy a company that any fool can manage—because eventually one will.
Peter Lynch

If you are wondering what happens to your tax dollars, here is a piece of late news: You just lost a measly half-billion dollars, which really isn't much for government work. It's really only 500 million dollars.

It seems that the Navy needed two new oilers, ships that gas up other ships. The Navy looked around for a ship builder and out of all the possibilities chose one nobody had ever heard of.

How could this happen? Well, it seems that Senator Arlen Specter of Pennsylvania wanted the ship built in Philadelphia because of unemployment there so he got into the act as senators are wont to do. The senator screamed, stomped, ranted, raved, cursed, discussed, endorsed, begged, pleaded, and threatened. Senators seem to have a special charm so the job went to the unheard-of outfit.

The Navy let two oiler contracts for 222 million dollars each and held its breath. When the money ran out, there were still no oilers. The company had bungled the job but pleaded with the Navy for more money. The Navy relented, but the company went broke anyway.

The Navy then moved the contract and the ships to George Steinbrenner (owner of the New York Yankees baseball team) and his company in Tampa. We don't know how the Yankees got mixed up in this. George ran out of money too and asked for more. The oilers were still screwed up. The Navy got mad and called it quits.

Result: a half-billion of your tax dollars later, we still have two non-oilers. They are resting quietly in mothballs. The Navy is paying $45,000 *PER DAY* for security! That's $1875 per hour! That will pay for 234 guards at $8 per hour for around the clock security. These are apparently very important oilers even if they don't oil. On weekends and holidays the security only goes up to $90,000 a day. This is a real bargain for taxpayers.

Surely this half-billion dollar fiasco calls for an investigation so we decided to do one. We went first to the fountain of all wisdom, the U.S. Senate, to see Senator Specter. After we forced our way into his office, he was very friendly and cooperative. He said, "Don't bother me, boy. Can't you see I'm busy running for president?" He really was.

Undaunted we went to the shipyard to talk to the company. We asked company president W. E. "Wee" Sanke if he had ever built a ship before.

"Hundreds," he said. "Nothing to it."

"Where can we see one?"

"At Toys 'R Us™," he said. "Kids love to play with them in the bathtub."

"What about the oilers?"

"Piece of cake," said Sanke. "An oiler is just a big oil can. We've made thousands of oil cans."

We were still getting nowhere so we went down to interview some workers at the dry dock which was under water. They were all sitting around playing cards, waiting for their next job.

"What about the oilers?" we asked Gomer Seaworthy, a likely-looking boat builder.

"They were just for practice," bragged Seaworthy. "Next time we're building an aircraft carrier."

"Oh my God," we cried. "Do you guys have any experience building ships?"

"Just these two oilers," Gomer said proudly. "Before that I was a barber. That guy over there is a chiropractor. We have plumbers, garbage collectors, sewer workers, cooks, stockbrokers, bill collectors, used car salesmen, ex-politicians, janitors, two lawyers, one Fuller Brush man, and one welder. How's that for experience?"

We noticed that the oilers were slowly sinking under the weight of 234 security guards. Senator Specter can be proud. This is a nice way for him to begin his presidential campaign. It must be comforting to you taxpayers to know that your shipbuilding is in good hands and that your tax dollars are working. Barely!

Now You See It, Now You Don't

> If the B-2 is invisible, just announce you've built 100 of them
> and don't build them.
> *Congressman John Kasich*
> *Chairman, House Budget Committee*

Congress is fighting the B-2 bomber battle in Washington right now. The question is whether to build any more of these invisible "stealth" airplanes. Naturally there are pros and cons.

Proponents say we are disarming too rapidly and getting too weak militarily. We won't be ready for the next war when it comes. There has been a war every 20 years throughout history starting with Adam and Eve.

Opponents say the BIG threat has gone away; the cold war is over. We don't need these expensive bombers for little brushfire wars. Furthermore, Congress doesn't like to pay for something it can't see.

The bomber has had problems. Its radar can't tell the difference between a rainstorm and a mountain. When it rains, the invisible bomber is visible on radar. It is not infallible.

The idea of an invisible airplane intrigued us so much that we went out to the base to take a look. We didn't know quite what to expect.

We called on Colonel Secrette who is in charge of invisible airplanes.

"We would like to see the B-2," we said.

"Whaddya mean see it? It's invisible," he emphasized.

"How do you fly it if you can't see it?" we asked, feeling stupid.

"You can't see it, but you can feel it," he assured us.

The Colonel blindfolded us and led us to the hangar where he turned us over to the guard, a Sergeant Pfeelwell. We shook hands with Pfeelwell, but we couldn't see him.

"Are you blindfolded?" we asked.

"Yes, sir. What do you want?"

"We would like to feel the B-2. Can't we take these blindfolds off?"

"That's impossible. We want to make sure you don't see anything. If it rains, you might be able to see it. But you can feel it," he promised.

"How do we find it?"

"We tried to use seeing-eye dogs, but they couldn't see it either. Now we use sniffer dogs. They can smell it. Here's your dog, Spot. Hold on, and he'll take you right to the airplane."

We followed Spot until we bumped into something solid. We felt all the way around whatever it was. It didn't feel anything like what we had read about in the newspapers, just a lot of sheet metal and what felt like a pair of big rubber tires. We came away mystified. We could tell that members of Congress were also there, feeling their way around. They were not happy.

Intelligence reports indicate that the Russians are developing a defensive satellite which can smell B-2's and then rain on them.

The Battle of Lexington

> This is not a conventional war. We have to forget propriety.
> *Colonel Robert A. Koob*

Plans for homeporting the aircraft carrier *USS Lexington* are up in the air. Pensacola wants the Lady Lex to stay permanently, either on active duty or later as a museum. We can now report our plan of attack.

First, we declare war on Texas. Texas wants to steal the Lex, but Texas is unsafe. It is far too vulnerable to attacks to recover the Alamo by Mexicans.

Next, we appoint Admiral Jon Weissman as supreme allied commander of the resurrected armies of the confederacy, known as the sons (and daughters) of Robert E. Lee. (Robert E. Lee appears in the New Testament.) The plot to steal the Lex is Yankee inspired. Teddy Kennedy is somehow involved. The South might have been better off if Paul Revere had fallen off his horse.

Supreme allied forces headquarters is located at Trader Jon's Downtown Social Club at 511 South Palafox Street.

The Pensacola wing of the confederate air force, consisting of a 1941 Stearman, a boomerang, and lighter-than-air party balloons, is under the command of Col. Ellis Davis. His deputy and navigator is a former air force major (equivalent to an able-seaman) with a distinguished combat record. This ancient WWII airman knocked out three bridges and an ammunition dump, and then they sent him overseas.

Admiral Weissman's battle plan is to sink the *Lexington* in place, then capture her with a boarding party commanded by South Palafox mayor Clark Thompson. The boarding party consists of charter members of the South Palafox Riff Raff and Procrastinator's Society armed with weed-killers and squirt guns.

The supreme allied commander has discovered some dehydrated confederate cannonballs behind his bar. The ordnance division, headed by attorney G. Brown, Esq., plans to rehydrate these munitions with Perrier and issue them to Col. Ellis Davis, who will then fly over and drop them on the Lex.

Should this fail, a secondary attack will be launched by Col. Earle Bowden commanding the Confederate artillery in Ferdinand Plaza. Col. Bowden has discovered that two of the cannons there point broadside at the carrier. The artillery can fire the cannon balls safely through the new cultural center by opening the windows. With time-delay fuses, this attack should succeed if all else fails.

Of course Yankee spies have notified Capt. Flack Logan of the Lex of this plan. The crew is on full alert and has redoubled its guard. Capt. Logan, in the immortal words of Gen. George Armstrong Custer, has issued his battle order: "Men (and women), we are completely surrounded by the enemy. Do not let a single one escape."

In a gesture of friendship and reconciliation, Capt. Logan has promised not to use torpedoes against Commodores John Fogg, Jerry Maygarden, and Vince Whibbs, commanding Confederate naval forces, who will be attacking from the sea in an encircling maneuver. Torpedoes would pose a grave threat to the Confederate navy, manned by City Council members and consisting of lifeboats, canoes, kayaks, rubber rafts, and two armored dreadnoughts dredged up out of Mobile Bay. We are reminded of Admiral Farragut's immortal words, "Damn, torpedoes,....etc." It is most fortunate that historic episode will not be repeated here.

As a friendly reciprocal gesture, Admiral Weissman is expected to be magnanimous in victory. He has promised that Capt. Logan may keep his sword and horse. Nor will the Captain be forced to walk the plank but will be cordially invited to surrender and remain in downtown Pensacola with free parking and a lifetime pass to Trader Jon's. But the Navy frowns on surrender.

In a stirring pre-battle address, the supreme allied commander rallied his troops and issued his famous battle order, "Do not fire until you see the whites of their eyes." We believe he may have borrowed this speech from somebody else, as we think we may have heard it somewhere before.

We have a war correspondent in an inner tube on the scene and will report further action as it develops.

We want the Lex! To arms!

Curtains

> *Yogi Berra: "You mean you get sea sick?"*
> *Rube Walker: "Do I ever!"*
> *Yogi Berra: "On water?"*

The following story is true. Not even the facts have been changed to protect the guilty. This is an actual recording of a phone conversation between Corpus Christi, Texas, and Pensacola, Florida. The phone call was intercepted by illegal means. The reporter is in jail for refusal to reveal his source...and for safekeeping.

Pensacola—Hello, Corpus. This is Pensacola calling. Is that you?

Corpus—Yes, this is Corpus. We're sitting here waiting for the *Lexington* to come in. It was nice of you to call and congratulate us. We're sorry you lost out. Too bad!

Pensacola—Too bad, my foot! We're calling to warn you. You've made a terrible mistake. You really screwed up. We studied that Lex for six months, and you've bought a loser. It's a lemon.

Corpus—Oh my gosh, don't tell us that. The Lex is coming in right now. We studied it for six minutes, and everything looked good so we went right ahead. What did you find out?

Pensacola—Well, you need to know the facts. That old tub leaks, she won't float, she's rusty, she's worn out, you can't maintain her. Nobody wants to look at that pile of junk. She'd be better off as razor blades.

Corpus—Good Lord, you don't mean it. We checked with Charleston and Mobile, and they said hundreds of thousands of tourists visit their ships every year. They even said they make money.

Pensacola—Lies, nothing but lies. They're a bunch of first-class liars. They're just covering up their own mistakes. In Charleston, the *Yorktown* got hit by a hurricane and tipped over. She's upside down. You have to climb up to the engine room and down to the bridge. You guys have really ripped it.

Corpus—Oh Lord, what a disaster. What have we done?

Pensacola—We really feel sorry for you poor people. We wish we could help, but it may be too late. We tried to warn you.

Corpus—Good God, we'd better get rid of this lemon. What'll we do? Do you want her back?

Pensacola—Heavens, no! But maybe it's still not too late. Just get the tugs to hook back up and haul her back to Philly for scrap.

Corpus—OK; good idea. Praise the Lord! I think you've saved us. I'll holler at the captain. (Noise of window opening and shouting. Then a gasp and long silence.) Oh, my God! The ship's gone! She sank! The captain went down with her. (Sound of gurgling and bubbles rising to the surface.)

Pensacola—Too bad. We're sorry. We tried to tell you. God rest her soul. May she rest in peace.

(The curtain falls, forever.)

Everybody Does It

> One of the evils of democracy is that you have to put up with
> a man you elect whether you want him or not.
> *Will Rogers*

Our country's morals are slipping again. Someone recently wrote that we're already in worse shape than Rome before the decline and fall.

We lead the world in illegitimate babies. Some aborigines have a better record than ours. Our role model is Hollywood where bragging rights go to Madonna and Jodie Foster who recently produced babies without benefit of matrimony. In Hollywood that's standard procedure.

We won't even mention drugs, AIDS, or presidential activities; the public doesn't seem to give a damn about lying, cheating, and adultery.

Now the armed services face a rash of adulteries all the way up to generals, not to mention sexual harassment and actual rape. The secretary of defense wants to ease up on punishing adulterers in the services. The Marines are raising hell about any attempt to reduce the penalty for adultery by service folks.

We wondered why the Marines are so outraged since the other services seem ready to comply. We interviewed veteran Wright Phace, a gnarled Marine gunnery sergeant.

"Why are you guys so hard on adultery? They say everybody is doing it," we said.

"We ain't everybody; we're Marines. We don't make love with other peoples' wives; we make war. We ain't got time to mess around with sex maniacs. Throw the bums out."

"You guys are awful tough. How about a little tolerance? The White House says, `Don't ask, don't tell.'"

"The Marines invented 'don't ask, don't tell.' If the grunts in boot camp don't ask why they have to crawl through an obstacle course with their noses in the mud while we're shooting over the tops of 'em, we damn sure ain't gonna tell 'em; and they'd better not ask. We just tellem 'do it,' no questions. But if I ask a question, they'd damn sure better answer. It don't work both ways."

"Is that all you tell them?"

"No, I tell 'em, 'Give your hearts to God 'cause your butts belong to me'."

"That's not exactly what the White House had in mind."

"The White House stole the idea from us in the first place. Now they're using it against Starr. If Starr don't ask, they don't tell. Hell, if Starr asks, they don't tell anyway."

"You'd better be careful. You're talking about your commander-in-chief here. We were talking about adultery in the Marines."

"Marines don't have time for adultery. We're too busy getting ready to fight. But if one of our little so-and-so's sneaks around with somebody else's wife, the Marines won't stand for it. If the husband don't kill him first, we will. He's outta here!"

"But certainly you'll follow orders from Washington."

"Yeah. I'm a good Marine. But if I could get my hands on the White House, they'd all go through boot camp. They'd shape up or ship out!"

He stomped out and chewed on a gang of recruits. Remember Dibble's First Law of Sociology: "Some do, some don't."

Gropings

> Ladies, without distinction of sex, would be welcome.
> *Political Handbill, circa 1900*

Sexual harassment is apparently here to stay.

Courts recognize all kinds of hanky-panky: off-color jokes around the water cooler, disparaging remarks, pinup calendars, glass ceilings, etc. Cadets at the Naval Academy chained a lady cadet to a urinal just for fun. The mean boys at the Citadel hazed several female new cadets right off campus, permanently.

Macho male bosses proposition lady employees with unveiled threats. Cooperate or else! Female bosses are known to have used the same tactics on male employees; turn about is fair play.

The poor military services seem to suffer the most. The U.S. Army just tried its Sergeant Major, Gene McKinney, for allegedly molesting female sergeants under his command. The women lost.

Remember Lt. Kelly Flynn, the lady B-52 bomber pilot? She bedded with the husband of an Air Force enlisted person. This is sexual harassment by proxy. Out went Lt. Flynn, clean out of the Air Force.

With all of this negative activity, the services are understandably paranoid about non-regulation sex. We recently heard of an army regulation which requires soldiers to stay at least 6 inches apart in all formations, including chow lines. We are not kidding about this.

This is an interesting rule. We are puzzled as to how the army could enforce such a regulation so we asked Private Ima Gonner, a brand new army recruit.

"What about the 6 inch rule?" we asked Gonner.

"That's the first thing they told us after they swore us in."

"How do they enforce it?"

"We all carry 6 inch rulers. If some creep gets closer than my ruler, I blow my whistle, squirt him with mace, and kick him."

"Is 6 inches enough?"

"No way. We need 6 feet. Some of these gropers have long arms. Nobody's safe at 6 inches."

"Have you been groped?"

"Daily. I wore out my whistle, and I'm outta mace. Fortunately my sergeant always saves me just in time."

This sounded very serious to us so we questioned Sgt. Justin Tyme, Gonner's platoon leader.

"Do you believe in joint basic training with males and females?"

"$@!?!* No! I've got some sex maniacs on my hands. I had to put chastity belts on the whole damned platoon, both male and female. The Marines have the right idea: keep 'em apart until their hormones get adjusted to military life. In the old days if they didn't behave, we just drowned 'em in a swamp. We can't do that any more. It's considered politically incorrect."

"Is the 6 inch rule working?"

"Hell, 6 inches don't help. I try to keep 'em at least 10 feet apart. I keep some of 'em in cages. I didn't join the army to run no finishing school for young ladies. I'm gonna retire." He spat disgustedly.

"What about the commander-in-chief?" we asked.

"We use a 100 yard tape measure for him."

Just then a whistle blew, and the sergeant rushed off to rescue another fair damsel in distress.

All's Fair in Love and War

> I am glad to say that the first man to knock him down for doing such a
> thing was his own wife.
> *Anonymous*

"We must 'ungender' the military services." So says a female Duke University professor whom the Army hired as a consultant on harassment of female solders.

She wants to eliminate "masculinist aggressivity" by bonding, using as models Alcoholics Anonymous and communist cells, among others, so help us! She is out to "feminize" the military services. We have no doubt she is right since the Army pays her.

Just to make sure we visited Fort Eleanor Roosevelt to get a first-hand look at how we will fight in the future. We met Colonel Herman Hanover riding in a pale pink convertible, which seemed a bit odd to us.

"What about this?" we asked.

"This is our latest combat personnel carrier. Isn't it divine?" he said.

The public affairs officerette next introduced us to Captain Wilhelm Wannabee who was wearing a skirt and high heels.

"Is that the new uniform?" we asked.

"Yes, one model for all, male and female. We're bonding, you know. Everything's for love."

"How is the combat training going?"

"It's so, so marvelous. We just love it." He sent us to see M/Sgt Muscles Malone, a burly, tough, no-nonsense, hair-on-his-chest drill sergeant. He was wearing a camouflaged chiffon sheath with matching accessories.

"What sort of uniform is that?" we asked.

"This is the latest combat gear. Isn't it lovely?" He led us to the hand grenade range where troops were practicing. They were throwing powder puffs at targets.

"What on earth are the soldiers doing here?" we wondered.

"My dear man, we don't have soldiers any more. These are all soldierettes, both male and female. We're bonding. We're all in this together. Everything's for love."

"Are these grenades effective?'

"My dear sir, we don't throw hand grenades any more. We're throwing tokens of love. Aren't they cute? We want to bond with the enemy."

"What if the enemy doesn't want to bond? Suppose they shoot back with real bullets?"

"Good gracious, we hope not. We want their soldierettes to love us. When the captain yells 'charge', we'll all rush out and hug each other and hold hands. Everything's for love."

"Can this crowd fight if they have to?"

"Heavens, no. We don't fight; we just love."

"What about the enemy?"

"We don't have enemies. We just have friends. We love everybody. Go in peace."

We got out of there as fast as we could before the sergeant could hug us. We hope the enemy gets the message in plenty of time.

In the meantime, we hope the captain never has to yell "charge." If he does, the lady Duke professor should be the first one out of the trench.

Peace! We hope.

ON POLITICS

A lot of my friends are politicians and I'd trust them anyplace ...
except in office.
Anonymous

There is no distinctly American criminal class except Congress.
Mark Twain

Politics takes a lot of money just to get beat with.
Will Rogers

The primary purpose of a political party is to throw their rascals out
and put our rascals in.
James Kilpatrick

I don't belong to any organized party. I'm a Democrat.
Will Rogers

Christopher Columbus was the world's greatest politician. When he left Spain,
he didn't know where he was going. When he got to America, he didn't know where
he was. When he got back to Spain, he didn't know where he had been.
And he did it all at government expense.
Unknown

The Royal Flush

The law of probable dispersal: That which hits the fan is not evenly distributed.
Chamberlain

The news media recently reported a serious incident which occurred at Pensacola
City Hall. The matter has been treated as a joke, but in reality it was no laughing matter.

During a hot and heavy debate on an important issue at the city council meeting,
the Honorable John Fogg, mayor of Pensacola, received an urgent call to nature. With
due apology he abandoned his post and retired hastily toward the men's room located
to the rear of the chamber.

As the debate continued, suddenly over the public address system came a loud,
long, satisfying cry of relief, "Ah...h...h...h...h!," followed by loud flushing.

Horrors! His honor's portable mike switch was in the full *ON* position, and he was
still plugged into the loudspeaker system throughout the building. The mayor was
broadcasting live to the entire audience.

The city attorney, who was also in charge of the men's room, rushed back to un-
plug the mayor. Turned off, but too late! Needless to say, his honor returned to the podium somewhat flushed.

His honor, still flushed, had the last word: "At least when we're back there we know exactly what we're doing."

Wet or Dry

> We are in favor of a law which absolutely prohibits the
> sale of liquor on Sunday, but we are against its enforcement.
> *Democratic platform in Syracuse, NY, 1920*

Wet or dry elections are still an old southern custom. The question pops up regularly and brings on civil war. "Whiskey" is a fighting word.

We were not sure where we should stand on this issue so we consulted one of our elected friends for advice.

He wrote us as follows:

"My Dear Friend:

"I had not intended to discuss this controversial subject at this particular time. However, I want you to know that I do not shirk controversy.

"On the contrary, I will take a stand on any issue at any time regardless of how fraught with controversy it may be.

"You have asked me how I feel about whiskey.

"Here is how I stand on this question:

"If, when you say whiskey, you mean the devil's brew, the poison scourge, the bloody monster that defiles innocence, dethrones reason, destroys the home, creates misery and poverty, yea literally takes the bread from the mouths of little children, if you mean the evil drink that topples the Christian man and woman from the pinnacles of righteous, gracious living into the bottomless pit of degradation and despair, shame and helplessness and hopelessness, then certainly I'm against it with all of my power.

"But if, when you say whiskey, you mean the oil of conversation, philosophic wine, the ale that is consumed when good fellows get together, that puts a song in their hearts and laughter on their lips and the warm glow of contentment in their eyes.

"If you mean Christmas cheer.

"If you mean the stimulating drink that puts the spring in the gentleman's step on a frosty morning.

"If you mean the drink that enables a man to magnify his joy and his happiness and to forget, if only for a little while, life's great tragedies and heartbreaks and sorrows.

"If you mean the drink the sale of which pours into our treasuries untold millions of dollars which are used to provide tender care for little crippled children, our blind, our deaf, our dumb, our pitiful aged and infirm, to build highways and hospitals and schools, then certainly I am in favor of it. This is my stand, and I will not compromise.

"Sincerely, Your Friend.

"P.S. I have also taken a firm stand on abortion."

Honesty In Politics

> Politics is an excellent career unless you get caught.
> *Robert Half*

Honesty is always the best policy but not necessarily in politics. In politics there are other ways. The end justifies the means.

Chicago is famous for politics. A little boy was standing on a Chicago street corner crying. A lady stopped and asked him, "What's the matter?"

"My father doesn't love me anymore," sobbed the youngster.

"Of course he does," she replied. "What makes you say that?"

"He's been back to vote three times since he died, but he's never come to see me once."

Jersey City is another unique political place. An investigator went to check on reported voting irregularities. He reported back that "he just found out why the Democrats always win in New Jersey."

"Why?" they asked.

"Because yesterday somebody broke into the mayor's office and found next year's election returns."

Individual politicians are not immune to persuasion and influence. You may have read recently about all the lobbyists in Florida, twice as many as any other state. They are also called legislative engineers in polite circles.

We may have the most lobbyists of all, but Texas, as always, has the biggest. Zeke Manners, a popular comedian, tells this story:

A lobbyist rides up to a Congressman's house in a brand new $90,000 Rolls-Royce. He tells the Congressman, "I want this car to be yours."

The Congressman says, "Common decency and morality would never permit me to accept a gift like that."

The lobbyist replies, "I can understand that, sir. Now suppose I sell you that car for $10.00."

The Congressman thinks about it and then says, "In that case, I'll take two."

New Yorkers have the biggest political messes of all, but they always come out smelling all right. If a dog messes up on the streets in New York City, the fine is $25.00; if a politician messes up, the city gets a $3 billion federal loan. I guess that's not dishonest; that's just politics.

In politics, there is some honor among thieves....but not so much.

Back from the Dead

> We must restore to Chicago all the good things it never had.
> *Richard Daley, Mayor*

Mike Royko of the *Chicago Tribune* is our favorite columnist. With great finesse and good humor he sticks it to the hypocrites and big shots.

He recently harpooned Ed Rollins and the Republicans in New Jersey for paying voters not to vote. Who ever heard of such nonsense? You are supposed to buy *votes*, not *unvotes*. Royko wrote that he never imagined that Republicans were smart enough to pull off a stunt like that or dumb enough to admit it. Rollins bragged about it, then said he lied about it. Worse than dumb! The Democrats are fighting mad about it; they want to get even.

After all, said Royko, Democrats have been cheating and buying elections for years. It's about time for the Republicans to wake up, get on the ball, and start winning some for themselves.

All is fair in love and war and politics. Two politicians were in the cemetery copying names off all the tombstones. When one suggested they also go to a second cemetery, the other wanted to know why. "Because those people over there are just as entitled to vote as these people over here."

Unfortunately some of the voters are dying out. The lesson here is to vote, dead or alive. One little old lady told us, "The reason I don't vote is that I don't want to be held responsible for what goes on in Washington, D.C."

No matter how you manage to get elected, it's not all gravy when you finally win. We recalled having been elected to the Florida House of Representatives, mostly by live voters, we think. Naturally we were extremely elated and retoured the district to thank our voters.

We stopped at one country house where an elderly lady was fanning herself and gently rocking on the front porch.

"Well, what did you think of our election?" we asked innocently.

"Well, to tell you the truth, sonny, I didn't even want a representative, and you're the nearest thing to nothin' I could find."

Flies, Fries, Fish, and Stumps

> The trouble with elections is that somebody has to be elected.
>
> *P.J. Rourke*
> Parliament of Whores

The good old days of southern Democrat politics have disappeared from our county.

Thirty-five years ago there was only a hardy handful of Republicans here. They met occasionally in a phone booth. Everything was Democratic.

If you went down to register as a Republican, the registrar's office had never heard of such a thing. Nobody knew what one looked like. The registrar snickered and looked at you funny and whispered something about communists.

TV rallies have replaced the good old-fashioned rootin'-tootin' rally attended by the Democratic Committee, the Democratic candidates (there were no Republican or communist candidates), all their families, friends, and supporters, and three or four other voters. You can tune out a TV rally, but you can't very easily sneak out of a real live outdoor rally.

In the good old political days the Republicans didn't have enough folks to hold a rally. Most remained in the closet. They couldn't vote in the primaries as they had no candidates. Winning in the Democratic primary was tantamount to election.

It's really too bad the old fashioned political rally, the press-the-flesh type with all the flies, fries, and fish, has disappeared. You can't really appreciate southern Democrat politics unless you have attended a real honest-to-goodness outdoor rally. We went to one once.

The affair began with a fried fish box lunch. The cost was designed to keep Republicans away, and it did although several poor Republicans did show up and spent the evening languishing outside the fence, eating their own peanut butter sandwiches which they brought from home.

One fat Democratic candidate for sheriff lost quite a few hundred votes when he barged ahead of the line waiting for fried fish. Hungry voters don't mind lining up as long as overfed candidates don't muscle in ahead of them. There were loud cries of protest and some shoving. A Republican finally felt sorry for this fat, hungry politician and gave him half a peanut butter sandwich.

The candidates spoke from a platform on the back of a dump truck. There was a microphone which didn't work, but the candidates were shouting like revival preachers and thus didn't need it. They were fired up.

The jovial Democratic chairman started things off after dinner. He spoke for twenty-three minutes and introduced everybody except the Republicans. The candidates were only allowed five minutes. Those were some of the longest five minutes we ever heard. The chairman pulled their shirttails after eight or nine minutes. After ten minutes he hit them with the gavel. Meanwhile, the flies were buzzing around the leftover fish and chewing on the crowd and the candidates.

One candidate, who had apparently enjoyed refreshments along with his box lunch, went to sleep and fell backward off the platform. The committee scooped him up quickly and propped him back up on the platform. When it came time to speak, he made more sense than the rest of them and took only three minutes.

According to the candidates now in office, the county has never been in better shape. According to the candidates trying to get into office, things have never been worse. It's hard to know whom to believe. Voters came away confused.

The Republicans got mad and asked for equal time but didn't get any. They tried to turn off the lights at 9 o'clock but failed.

Finally the owner of the dump truck got tired and wanted his truck back. He pulled the release lever and dumped the whole proceedings onto the ground. When the lights went out, the chairman adjourned the meeting, and everybody went home in the dark.

One other great feature of the good old Democratic rally days was the stump speech on the campaign trail. The candidate actually stood on a stump (or a 55 gallon drum if there were no stumps) and held forth after the warm-up country music stopped.

We heard one candidate speak to an Indian tribe on the reservation. Standing on his stump, he promised them everything, including the moon.

He promised better housing. The crowd shouted "Oom Gallo Gallo."

He promised better jobs. "Oom Gallo Gallo" they roared.

He promised better health care. "Oom Gallo Gallo" resounded.

When the candidate had finished, he noticed a prize Indian bull in the nearby pasture. He asked the chief if he could go out in the pasture to get a closer look at this beautiful animal.

"Go right ahead," said the chief, "but while you're out there, be careful not to step in the Oom Gallo Gallo."

Al Who?

> If you always tell the truth, you don't have to remember anything.
> *Mark Twain*

Columnist Tony Snow calls them all liars. Congressman Joe Scarborough says it's a case of mass amnesia. Whatever it is, it's downright serious.

Memory loss has infected the White House, its staff, and the Democratic National Committee; they can barely remember their own names. Congress is investigating fund raising during the last presidential election. Special prosecutor Kenneth Starr is checking up on the whole Whitewater mess. Suddenly a lot of minds have gone completely blank.

It all started with Presidents Bill and Hillary Clinton. At first Bill can't remember Paula Jones. Now he doesn't remember making any fund raising calls from his office, and nobody remembers getting any.

Bill and Hillary don't remember much about Whitewater except that is was a creek some where. Attorney Hillary doesn't know how her legal billing records disappeared for years only suddenly to reappear in the study in the White House. Nobody can remember who hired the White House security chief who ended up with a lot of FBI files that nobody can remember asking for.

It gets more and more forgetful: Vice-president Al Gore can't remember raising campaign funds in a Buddhist temple in California. First he remembered that it was just a "community outreach" meeting; then he remembered that it was "finance related." Finally he remembered that it was really "donor maintenance"—whatever that means.

The donors were monks and nuns who had taken the oath of poverty. Their contributions were paid for by others. They admitted they shredded the evidence, and they were given immunity in order to testify.

Al is trying to forget the whole thing, but Congress remembers. He forgot about a number of illegal fund-raising phone calls from his White House office. This is a no-no, but all he could remember was that there was no "controlling legal authority" on the question—whatever that means. He has now remembered to hire two lawyers to defend him.

Clinton friends Jim and Susan McDougal didn't want to remember much about the Whitewater land deal. Susan is in jail on fraud and contempt charges, and she won't talk. She may remember too much. Jim started remembering in turn for a reduced sentence. He also resides in jail. He recently died.

Web Hubbell, another Clinton friend and law partner, served as their deputy attorney general until he got caught. He defrauded clients at Hillary's law firm and went to jail. Then he apparently started remembering too much about Whitewater so he took the Fifth Amendment. That's a convenient form of amnesia.

Now comes Don Fowler who was the Chairman of the Democratic National Committee (DNC) during the last election. The committee raised money, switched funds around from soft to hard, strong-armed supporters and the National Security Council, got crooks into the White House, took foreign money, and generally performed miracles. Everybody at DNC remembers except poor Don.

Don tried hard, day and night, he said, to remember, but memory simply failed him. Prodded with memos and other testimony about what he did and said, Don still drew a lot of blanks. When asked if President Clinton was involved, he asked, "President who?"

This mental illness could get really serious. Suppose nobody can remember who has the keys to the Treasury? We think the problem must be in the water at the White House. Somebody needs to check out the plumbing.

In the meantime, when asked if Hillary had been involved, the President said, "Hillary who?"

Inside Africa

Talk does not cook rice.
Chinese Proverb

Our President Clinton has returned from visiting Africa for two weeks. In addition to apologizing for Americans, he drove around in the bush where wild animals roam. He chased elephants which reminded him of Republicans. He was accompanied by numerous press, staff, and camp followers. He thoroughly enjoyed himself.

On one of his drop-in visits to a small village in the middle of nowhere, his helicopter landed and blew the roofs off of two native houses. The blast rearranged the furniture and generally wreaked havoc. The inhabitants, who were blown about considerably, were not pleased.

One home owner rushed out to confront our president. He carried his spear, which alarmed the Secret Service. His name was Mfume Mfume. We think this is pronounced Mfume Mfume, but we're not sure. He spoke in Swahili, but we have done our best to translate his remarks.

"What the hell you do to my poor house?" wailed Mfume.

"I'm sorry, but we're from the U.S. Government, and we're here to help you," advised our president.

"Me got no roof."

"What's your name, young man?"

"Me Mfume Mfume, and me mad as hell."

"How do you pronounce your name?"

"Em-foo!-mee Em-foo!-mee," shouted Emfoomee, obviously agitated.

"I've come to apologize to your village. What can I do to help?"

"Where my roof?"

"The U.S. has a disaster relief program called FEMA. They'll help."

"Me no want no damn FEMA. They just bureaucrats. Take too long. Me need roof now."

"How about some foreign aid?"

"Congress too tight. No pay U.N., either."

"Well, I can put you on Social Security and Medicare."

"They all broke. Nuts to that!"

Our president was clearly getting exasperated. "How about our welfare program? Then you can get a job."

"No jobs here."

"But we'll help you find a job."

"Me no work. Me hunt. Wife work. Wife no need welfare. Where my roof?"

Our president spoke to an aide. "Can you find this guy some beads? Maybe that will help."

"Me no want no damn beads. Me want roof."

"What on earth can we do to make you happy?"

"You got cigarettes?"

"No, no, no. We're opposed to cigarettes. We're adding $1.10 tax per pack."

"No smoke, no roof, no nothing. U.S. no damn good."

In desperation, our president asked, "What about a girl friend? I think we can fix you up."

"Now you talkin', boss. You fix. Don't need no roof. No tell wife."

Unfortunately, Mfume's wife came outside to look for the roof just in time to hear this last exchange. She hit him on the head with her cooking pot and hauled him by one ear back into the house, roof or no roof.

Mmfume was last seen walking around with his head in the cooking pot. As the helicopter took off, it blew the roofs off two more houses. The president said he was sorry, again.

The Body

> The president doesn't want any yes-men and yes-women around him.
> When he says no, we all say no.
> *Elizabeth Dole*

Professional wrestler, Navy Seal, decorated Vietnam vet, high school football coach, radio talk showman, bodyguard, rock singer, suburban mayor: you have just met Jesse "The Body" Ventura, the new governor of Minnesota. He looks more like famed Minnesota fullback Bronco Nagurski than a government issue governor.

Jesse ran with little money and very little platform except blunt talk. He will refund $1,000 per taxpayer, says he. He favors gay rights and might consider legalizing prostitution.

All that aside, Jesse now has to govern. He was a Reform Party candidate (remember Ross Perot?) with a Democratic state senate and Republican house. There are no reform legislators, and as yet he has no cabinet.

When asked where he would find department heads, etc., Jesse didn't know. One refreshing thing about Jesse "the politician" is that when he doesn't know, he actually says he doesn't know. His opponents laughed at such politically incorrect behavior, but it apparently endeared him to the young and the blue-collar workers who are used to lots of politically correct baloney.

He beat the likes of Humphreys, Mondales, and Freemans who are the fabled and perennial politicians in Minnesota. Nothing elite about Jesse who is shedding "The Body" image. Henceforth, Jesse says, he's "The Mind."

Jesse cuts news conferences short when agitated. A reporter recently asked him about coming up with a new state budget. "Oh, sheesh," said Jesse. That ended that, but it does not end the question as to where Jesse will take Minnesota.

We tried to imagine Jesse's inaugural address to be held at Joe's Gym. Following his inaugural, Jesse appears at the first official state news conference ever held in a gymnasium. Things might go something like this:

"Governor, are you going to govern from the ring?" asks a reporter.

"Yep, I'm right at home here," says the governor, flexing his biceps.

"Could you tell us where you will find $1,000 for every taxpayer?"

"Thank you, young man. That's an excellent question. Let me shake your hand."

Jesse crushes the reporter's right metacarpus.

"What about your cabinet?" asks another intrepid questioner.

"That's another excellent question." Jesse wrenches the reporter's right arm and severs the clavicle.

"How about your budget?" bravely ventures a third reporter. Jesse body-slams the questioner and tosses him over the ropes into the third row of reserved seats.

A fourth reporter dares to ask, hopefully, "Will we have legal prostitutes?"

"Oh, sheesh," says Jesse who turns off the lights and throws the whole press corps out into a snowdrift.

Minnesota is in for some fun, some body-slams, and some broken bones. Don't laugh. It just might work.

ON RELIGION

Sign on church: "There ain't no hell." Sign on church across the street: "The hell there ain't."

Let us endeavor so to live that when we come to die even the undertaker will be sorry.
Mark Twain

I am ready to meet my maker; whether my maker is prepaid for the ordeal of meeting me is another matter.
Winston Churchill

When the white man came, we had the land and they had the bibles; now they have the land, and we have the Bibles.
Chief Dan George

I don't like to commit myself about Heaven and Hell, you see. I have friends in both places.
Mark Twain

Minis and Quickies

The natives finally understood Christianity. They ate the missionaries.
Mark Twain

We have a new messiah among us—right here locally. The Reverend Thaine Ford and the First American Baptist Church have just invented the 22-minute church mini-service, which includes the 8-minute quickie sermon. This revelation is equivalent to the second coming in history.

The Reverend Ford's breakthrough in theology has already gained us worldwide renown. Nowhere in scripture or standard liturgical procedure was such a breakthrough fore-ordained. The Bible does not specify the length of sermons or services, but the standard for the industry is one hour plus any overtime.

For example, the Reverend Dr. James Pleitz, formerly of Pensacola, Florida, and later of Dallas, Texas, always conducted the one-hour standard service which ended promptly at 12:00 noon. No overtime, "no souls saved after 12:00," proclaimed the good Doctor Pleitz. He also said his congregation got up and walked out promptly at 12:00 noon whether he was finished or not. Dr. Pleitz was always a nervous wreck trying to meet the deadline before the whole congregation disappeared in a stampede to the door.

Somehow this fact must have registered with the good Reverend Ford. The mini-service should overcome this tension. "I'll be able to beat them to the front door," proclaimed Mr. Ford.

What do you get in 22 minutes? The opening, prayers, two hymns, the 8-minute sermon, offering, responsive reading, closing, everything — everything, that is, except communion.

No communion poses some difficulty for some communicants. They have solved this one omission by retiring to a nearby lounge after the service. This could cause future problems for attendance. Some communicants might be inclined to go directly to communion. (Trader Jon's bar is now considering starting his own mini-service with communion.)

Worshippers must remain alert and awake during the mini-service. There is no time for even a short nap. The collection plate passes very rapidly; it goes by like a frisbee. If you blink twice, it's gone, which might encourage habitual fumblers and also reduce collections as the plate sails by too fast.

As in all churches, there is the temptation for everybody to sit in the back of the church for obvious reasons. The good reverend stopped that practice in a hurry. He now passes the collection plate starting in the back. Now everybody wants to sit in front.

Worshippers have to be quick. One lady sneezed three times and missed the entire sermon. Those persons accustomed to being fashionably late will miss the entire service.

The 22-minute service will soon appear in the Guiness World Book of Records. We expect it won't be long before the competition will challenge this record. Several rival factions are already trying.

The speed-talking salesman in the Federal Express ads has squashed a full one-hour service into 15 minutes by talking fast. One church offers a catered service. An ingenious pastor has instituted a drive-in window for fast service. The collection goes into a tollgate type receptacle. There will be no end of ambitious counterfeiters trying to horn in.

The good Reverend Ford maintains God is on his side in this venture; we believe him. His popularity is growing like wildfire; he may have to move his services to the Civic Center. He claims his mini-service isn't for everybody. Don't bet on it.

There are, however, some inherent dangers in all of this. Although increased productivity is a material goal, it could have serious consequences for the clergy. A proliferation of short services might very well cause an oversupply of preachers. Their unemployment rates might soar. It might well come to survival of the fittest.... and shortest winded.

We are not at all sure how all of this will play out. Might not the standard 11:00-12:00 service disappear all together? It might be necessary to build more golf courses which would, of course, be a boost to the economy, but we might need to build fewer churches. There are many imponderables here.

In the meantime, the Reverend Ford's First American Baptist Church enjoys national notoriety. The reverend has already appeared in national media. He is famous overnight. He has even received an inquiry from the Pope and has faxed to the Pope an outline of the drill for a mini-service.

Sara Duncan once pleaded, "If you have anything to tell me, for God's sake please begin at the end." That is precisely what Mr. Ford has done.

In *Roughing It*, Mark Twain wrote, "Providence don't fire no blank cartridges, boys." The Reverend Ford is fully loaded.

No souls saved after 22 minutes.

Miracle of miracles! Will wonders never cease?

(P.S. We later interviewed our own preacher who declined to institute the mini-service. He felt his congregation wouldn't be able to stand the shock if he quit after 22 minutes. There was danger of a mass heart attack, he said.)

To Sleep, Perchance to Dream

Lead us in a few words of silent prayer.
Coach Bill Peterson

The polls on statistics, murders, abuse, fraud, and other social ills all seem to indicate our morals have gone to hell in this country. Our preachers are all out laboring in the vineyards and trying desperately to save souls, but matters just seem to be getting worse in spite of their good works.

The good brothers in frocks are growing discouraged. The numbers of people who profess religion are growing; those who practice it seem to be diminishing. Church attendance is down in some quarters; collections are down.

You will recall the good French vicar who threatened to charge admission to his church because revenue fell off. When he added a cover charge, his flock deserted him. We had warned him about that possibility by fax and advised him to continue free admission, get his flock inside, lock the doors, and then charge them to get out. That didn't work either.

The plight of our reverend friends is frightening. Now comes the pitiful news, straight from the Associated Press, of a preacher whose entire congregation kept falling fast asleep during his services. The good reverend was naturally demoralized; he couldn't figure out why he was failing in his divine mission. He souped up his sermons, then he shortened them. He polled the congregation for answers, to no avail. Nothing helped. He then called in medical experts to test for possible physical ailments, drugs, plague, etc. No luck! Some of his parishioners were actually comatose by the end of his sermons. He was now frantic.

About this time, and just by chance, the gas man happened by. He sniffed around, donned his gas mask, and sounded the alarm. Sure enough, gas was leaking into the church from a faulty heating system. The gas company repaired the valve, and the case of the comatose congregation was solved. Everybody woke up again just in time to hear the sermon. As Senator S.I. Hayakawa explained, "The best cure for insomnia is to get a lot of sleep."

Eternity—When Will It End?

I don't want to achieve immortality through my work. I want to achieve immortality through not dying. I'm not afraid to die.
I just don't want to be there when it happens.
Woody Allen

We receive all manner of solicitations over the phone. There is no end to them, especially while we're eating dinner.

We just recently received a call from our "friendly undertakers" at a local funeral home. They were selling a layaway plan. This is a pre-paid funeral complete with vault and all the trimmings, sometime in the far distant future, we pray.

Their ad says that this pre-planning (and pre-payment!) is "for the benefit of loved ones left behind." We thought to ourselves, this is also for the benefit of the funeral home left behind. We could just picture Digby O'Dell, that famous old undertaker, licking his chops at the thought of this investment in his future.

To add insult to this annoyance, we next received a notice of our 50th class reunion together with a gloomy forecast from our class mortician. "Some of you won't make it so be sure to send in your obituaries early," the announcement said.

Our life suddenly began to flash in front of our eyes. It occurred to us that we had better get our affairs in order. We wanted to leave a legacy behind, but right offhand we couldn't think of one.

We did manage to pull together some of the wisdom which somehow has gotten us to where we are today. These are the great and universal truths which men and women live by. The least we can do is to pass these along to you before we go. We hope it's not too late:

Age - Oliver Wendell Holmes at age 90 on passing a pretty girl, "Oh, just to be 70 again."
Planning - "It wasn't raining when Noah built the ark." Unknown
Future - "The best thing about the future is that it only comes one day at a time." Abraham Lincoln
Time - "No wonder time flies. So many people are out to kill it." Anonymous
Error - "To err is human, but it feels divine." Mae West
Pessimism - "A Puritan is a person who has a desperate fear that somewhere, somehow, there's a person who is happy." H.L. Mencken
Work - "I do not like work even when somebody else does it." Mark Twain
Wealth - "Lack of money is the root of all evil." Mark Twain
Wisdom - "Pain makes man think; thought makes man wise; wisdom makes life endurable." *Teahouse of the August Moon*
Truth - "If you tell the truth, you don't have to remember anything." Mark Twain
Trust - "Trust in God, but tie your camel tight." Persian Proverb
Ambition - "I always wanted to be somebody, but now I realize I should have been more specific." Lily Tomlin
Peace - "Regiments shall bear as their battle honors the names of wars averted." British Army
Eternity - "Eternity is two people and a leftover turkey." Jim Dent
Death - "Oh, Lord, give me something to die for. For if I have nothing to die for, I have nothing to live for." Old Chinese Proverb

Those are the great and eternal truths that we should all live by, or try to.

Life is a blast; life is also a terminal illness. Your friends at the funeral home are standing by to help.

Hang on as long as you can.

To Each His Own

Christ is coming back, and boy is he mad.
Anonymous

The Bible is a monumental piece of work. To some it provides the very foundation for life; every word is gospel to be taken literally. To others, not so literal in their interpretation, it is a guide, not always to be taken word for word. To still others it is a great work of art, an ageless masterpiece of literature.

In other words, it means different things to different people. Religion is a very personal thing. Personally, we think God has a sense of humor, or he wouldn't put up with a lot of humanity's nonsense. Just for fun we rounded up some different observations from an assortment of people with their own ideas:

I believe that if we introduced the Lord's Prayer in the Senate, the senators would propose a number of amendments to it.
Henry Wilson

When the white man came, we had the land and they had the Bibles. Now they have the land, and we have the Bibles.
Chief Dan George

If only God would give me some clear sign. Like making a deposit in my name in a Swiss bank account.
Woody Allen

God made water; it takes man to make wine.
Victor Hugo

God gives us our relatives; thank God we can choose our friends
Ethel Merman

All the things I really like to do are either illegal, immoral, or fattening.
Alexander Woollcott

Adam and Even had many advantages, but the principal one was that they escaped teething.
Mark Twain

Why did the U.S. succeed? South American explorers came seeking gold. The founders of the U.S. came seeking God.
South American President

He who despiseth trifles shall perish little by little.
Apocrypha

The natives finally understand Christianity. They ate the missionaries.
Mark Twain

God is on the side of the heavier battalions.
Victor Hugo

So to each his own. Religion can be both serious and fun. Fortunately in our country everyone has the freedom to worship in his or her own way. Sometimes we need to lighten up. For instance, one dyed-in-the-wool southerner wanted to know whether General Robert E. Lee appeared in the Old or New Testament.

There are a few charlatans out there who try to sell the Bible like snake oil. God forbid! Ours is to reason why on our own in our own way. For most believers the spirit can move you without your falling down or speaking in tongues or passing out. If that's your religion, fine. Go for it! But beware of so-called religious leaders who are fakes and frauds and hypocrites, God's purely self-anointed, exclusive practitioners. They will be found out sooner or later. Hopefully sooner.

As the Good Book says, "Beware of false prophets."

Keep *your* faith.

All God's Children

The best thing about the ten commandments is that there are only ten.
H.L. Mencken

Preachers can tell a lot of great stories about themselves. They also tell some great ones about their congregations.

One particular church needed more light. A member of the flock suggested a chandelier. Another member spoke up to protest. "That's not what we need. In the first place, nobody can spell it. In the second place, nobody can play it. And in the third place, what this church needs is more light." So let there be light.

Another church had a leaking roof and badly needed money to fix it. From the pulpit the preacher pleaded with his flock to contribute, but nobody responded. Finally, in desperation, he directly addressed Mr. Smith who was sitting in the front pew.

"Mr. Smith, you're a good family man and a good and faithful church member. Your business has been good, and you've been very successful. Surely you could help us get a new roof."

Mr. Smith stood up and addressed the rector. "Thank you for those kind words, preacher. I would like very much to help, but I've got problems of my own. My wife needs an operation, my mother needs to go to a nursing home, my son needs to go to college, and my father needs a heart transplant. If I can say no to all of those people, I can certainly say no to you."

A farmer in Texas took over a large plot of very poor, rocky soil and turned the bad land into a magnificent, productive farm. He was very proud of his hard work and his great accomplishment and invited his clergyman out to view his handiwork.

The preacher was delighted and amazed. "This is God's work," extolled the good rector.

"Well, I'll tell you what," said the farmer, "that may be true, but I wish you could have been here to see it when God had it by himself."

A more serious situation occurred when an escaped felon took on a disguise as an itinerant evangelist and began preaching from church to church.

At his next revival meeting he had just mounted the pulpit to begin his sermon when he was dismayed to see his former prison cellmate sitting in the front row. What'll I do now, he thought; what if he spills the beans?

But he quickly regained his composure and began again, "Brethren, the text for today's sermon is First Corinthians, verse one: He who seeth me and knoweth me, please say nothing now and see me after the meeting."

Elsewhere a group of friends was having a friendly contest to see who could squeeze the last drop out of a lemon. They squeezed and squeezed and squeezed until they were exhausted, but no one seemed to be able to squeeze it out. Finally the last man up took a firm grip and squeezed mightily. Out came the last drop.

His friends were amazed. "How on earth did you do that?" they all wondered.

"Well, to tell you the truth, there was really nothing to it. You see, I'm the finance chairman at the First Baptist Church."

It has been laughingly said that the only thing two Baptists can agree on is how much the third one should give. That's true for the rest of us as well.

Amen Corner

It is not best when we use our morals on weekdays;
it gets them out of repair for Sundays.
Mark Twain

We have two rectors at our church with a great sense of humor. Preachers without humor can be deadly. We love stories about preachers from preachers who can laugh at themselves.

For instance, a large group of clergymen attended a convention at a large hotel. In the adjoining room a number of salesmen were also meeting.

At lunch for each group the hotel served a cup of watermelon for dessert. The chef served the clergy plain watermelon; he spiked the salesmen's dessert with liquor. By mistake the waiters mixed up the orders and served the spiked watermelon to the clergy.

The chef was embarrassed and horrified. He finally mustered up enough courage to ask a waiter if the clergy were eating the watermelon.

"Oh, yes sir," said the waiter. "They're gobbling it up. They asked for seconds, and they're saving the seeds."

There are always stories about three clergymen: a Protestant, a priest, and a rabbi. In this case, the three were out in a boat fishing. They all got thirsty for a coke so the Protestant volunteered to go get some. He stepped out of the boat and walked across the water to shore, then back. The rabbi was impressed.

Next they needed some more bait, and the priest offered to go. He walked across the water to shore and back. The rabbi was highly impressed. He said, "Here, let me try that." He stepped out of the boat and sank in ten feet of water. The other two hauled him back into the boat. The priest felt sorry for the rabbi and remarked to the Protestant preacher, "Maybe we ought to show him where the stumps are."

The same three later discussed how they divided up the collection after their services.

Said the rabbi, "I draw a circle on the floor; then I toss the collection up in the air. Everything that falls inside the circle goes to the synagogue; everything outside the circle belongs to me.

Said the priest, "I draw a line on the ground, then I throw the collection up in the air. What falls on the right side of the line goes to the church; I get everything on the left side."

The Protestant minister was not impressed. "I've got a much better system than that: I toss the collection up in the air. Whatever sticks to the ceiling belongs to the church; what falls back down belongs to me."

Finally, a fan at the race track noticed a Catholic priest performing a brief ceremony for one particular horse before each race. Those same horses won the first three races. The fan figured he was on to something so he bet on the 4th horse to which the priest administered to win. The horse dropped dead on the back stretch.

The bettor was naturally upset and asked the priest, "How come?"

"Well," said the priest, "If you were a Catholic, you would know the difference between a blessing and the last rites."

Preachers are also fun, good sports, and very necessary.

The Joke Is On Us

> There will be a procession next Sunday afternoon on the ground
> of the church; but if it rains in the afternoon,
> the procession will take place in the morning.
> *A preacher to his congregation*

The world is losing its sense of humor. We have gotten much too serious. We have lost the touch of class of old-time humorists like Mark Twain and Will Rogers. Not much clean humor these days. On TV it's called comedy, which means four-letter words and filthy jokes.

There is nothing more gratifying than a clergyman with a sense of humor. We had just such a one as our guest preacher last week. Our regular young clergyman was a protege of this older rector so our guest started out by saying he knew that if he made any mistakes during the service, he would be sure to hear about it from his young friend who had studied and served under him.

Our guest preacher couldn't resist pointing out that our regular young rector likes to change things for the sake of change. His motto - "If it ain't broke, fix it."

Then our guest told us a story about a graveside service which he was called upon to perform. When he arrived at the cemetery, it was raining cats and dogs He pulled out his umbrella.

To his chagrin the umbrella was a loud red and white striped affair, and there in large letters was spelled out "Budweiser." The widow said her husband would have loved it. Our congregation did.

He went on to tell about a family during a violent thunderstorm. After they all went to bed, the little daughter called out from her room that she was scared. She wanted one of her parents to come and stay with her. Her father called back to comfort her, "You'll be just fine. God is in there with you."

A few minutes later she called out again for her parents to come. Again her father reminded her that God was right there with her. "Yes, daddy, I know," said the little girl, "but I want somebody in here with more skin on him."

Our guest was basically a small town preacher. He told of his parishioner Miss Helen, the church's musician. When the bishop was due to visit, the clergyman asked Miss Helen to prepare some special musical selections in the bishop's honor.

At the appointed time during the service Miss Helen shouldered her violin and rendered, of all things, the "Camp Town Races." The bishop was horrified; he rolled his eyes heavenward. The rector rolled his eyes at Miss Helen and signaled her to play her second number which surely would be something more appropriate. Out came "Danny Boy." The bishop was not amused.

The ladies of his church refused to take up the collection; that was a man's chore, they said. So one day when only the ladies and one unfamiliar male visitor were present at the service, the ladies prevailed upon this lone male to pass the plate. He finally reluctantly agreed.

He finally finished his chore in good order and, completely satisfied with his effort, proudly presented himself back at the front of the church with the collection plate which he delivered ceremoniously to Miss Helen. She then tapped him on the chest with the bow of her violin and said loudly, "Go!" He has never been back.

People in small churches must have more fun.

ON SCIENCE

Automatic means you can't repair it yourself.
Unknown

Sure, chlordane is going to kill a lot of people,
but they may be dying of something else anyway.
Othal Brand, Texas Pesticide Board

It works better if you plug it in."
Sattinger's Law

My grandfather invented the burglar alarm but never got credit.
Somebody stole it before he told anybody about it.
He also invented 1-up to 6-up, but then he stopped.
Victor Borge

Up, Up, and Awaaay

I think that the free-enterprise system is absolutely too
important to be left to the voluntary action of the market place.
Congressman Richard Kelley

Wonders never cease. The astronauts recently took off and flew into space with a new experimental $23,000,000 toilet. Twenty-three million. Believe it! That's 23 followed by six zeros. At one time a $700 aircraft toilet seat was a national scandal, but the ante has gone up considerably.

Obviously toilets are a problem in space, what with no gravity and everything floating around every which way. So the first problem with any johnny is to anchor it down.

The second problem is that the astronauts forget to keep the lid down. This is a capital offense in our family.

In spite of these problems, $23,000,000 still seemed a little steep so we went to Mosquito Flat to visit Herman Hobbgood, veteran mechanical genius and retired sawmill mechanic. We wanted some straight answers.

After we broke the news to Herman about the astronauts' new toilet, we asked, "Does this privy sound a little expensive to you?"

Herman scratched his head. "Well, I ain't never been in space, but we never had no trouble like that in the sawmill." Then we explained gravity.

"Well," said Herman, "that sure do put a little different light on the subject."

"Do you think you could come up with something competitive?" we asked.

"I'll think on it," he promised.

Herman called us back later and said he had the plans ready. We rushed over. What Herman has devised is a stroke of genius.

The answer is a portable pressurized outhouse with a full moon on the door. It has a standard Navy anchor to hold it in place. Inside is a set of ordinary bicycle pedals.

"Are those pedals to drive it around with?" we asked.

"Naw. On TV I seen them astronauts pedaling all the time for exercise. So now they can pedal inside here and make themselves useful. The pedals are hooked to the generator, the air compressor, and the fan. When you pedal, the generator energizes the magnets, which provide the gravity inside. The air compressor keeps it pressurized, and the fan is for exhaust."

"What about this door spring here?" we questioned.

"That's to keep the lid down," he said. Herman thought of everything.

There is also an artificial horizon so you can tell whether you're upside down. Herman pointed out the navigation lights so that the astronauts can take their commode outside with them during space walks and not lose it or have somebody run into it.

We asked him how much he thought it would cost to manufacture this rig.

He figured a minute and then said, "$23. It's all made from scrap and secondhand parts we found in the junk yard."

We have taken out a patent with Herman and have notified NASA. We expect a favorable reply.

Rain with Snow

> Everybody complains about the weather but nobody does anything about it.
> *Mark Twain*

Twain was wrong. Here's the true story of a lady who *did* do something about it. *She is suing the weather bureau.*

It seems that the lady's weatherperson failed to predict a cold wave. As a result the nice lady caught cold, lost four days of work, bought medicine, and suffered severe stress during her sniffles. She is suing for $1,000, a trifle in today's litigious society.

We visited Abe E.S. Corpis, a local attorney who advertises in the phone book, on television and radio, in the newspaper, and on milk cartons. His TV ads generally appear between the *Depends* undergarment ads and the deodorant commercials.

"What about this lawsuit?" we asked Abe.

"Ridiculous!"

"Why so?"

"Nobody sues for $1,000. There's nothing in it for the lawyer."

"What would *you* do?"

"I'd go for the big bucks. A million or two for compensation and a million or two for punitive damages. There's a lot of pain and suffering with a head cold."

"But she only had a cold. It wasn't terminal."

"Doesn't matter. Do you know what hospitals charge for aspirin? No? Well, I'll tell ya. The last aspirin tablet I bought in a hospital cost $100. That's heavy, man. Ten aspirin, a thousand bucks."

"What will you get out of this case?"

"Not much after my expenses. There may be a little left over for the little lady."

"Will you take the case?"

Abe had a gleam in his eye, and we could tell the little old lady's case was not the farthest thing from his mind. As we left, he was dancing a little jig.

Then we called the weather bureau and spoke to a Miss Stormi Wether, the meteorologist on duty. We asked if she had heard about the case.

"Absolutely. We're in total panic down here."

"Why so?"

"Why not? We can't afford to be sued over every little head cold. Where would it stop? If Corpis takes the case, we're dead!"

"What will you do to cover your you know what?"

"We just called Feran Hite, our boss in Washington, to find out."

"What did he say?"

"He said to cover our you-know-whats."

"How are you going to cover them?"

"Well, we'll have to beef up our weather reports. For instance, this afternoon I'm calling for a hurricane or two, some tornadoes, freezing rain, snow, fog, ice, etc. It'll be a rough afternoon."

"Good God, all that's coming at once?"

"Absolutely. I'm not making any mistakes if I can help it. We can't take any more chances."

"Suppose we have a heatwave? Then what?"

"No problem. I'm also calling for clear, partly cloudy, and cloudy weather with a drought, hail, extremely high temperatures, heat index at 150°, etc. Let's see Corpis sue us on that forecast. We're covered."

We could just see Corpis' little old lady client wearing her raincoat over her fur coat over her bathing suit. Thanks to Corpis and his client, forecasts will be much more complete from now on.

Nor Wind Nor Rain Nor Sleet Nor Snow...

There is some possibility of showers tonight although it is
probable there will be no rain.
Weather Report in the L.A. Express

The government weather service has purchased 350 million dollars worth of weather radar so that the weather stations will know what is going on outside their offices.

Now the government has fired 700 weather observers nationwide. These were the people who occasionally walked outside or looked out the window to see what the weather was really doing, not just what their charts said. They didn't trust their instruments. Eliminating these observers saved the government $25,000,000.

Now problems have arisen with the new radar. It seems that the radar can't tell the difference between rain or snow or fog or hail or ice or what have you. So now it will cost the same government $30,000,000 to train new observers.

Could our beloved government really have done such a thing? We went out to the weather station to find out. We met with Charles H. "Hurry" Cane, station chief, who referred us to his assistant, B.L.O. Hard.

"Could we see this new radar?" we asked.

"Sure," said Hard, "I'll take you down where we have it installed."

We went down a steep ladder into what looked like the catacombs. There we met Hugh "Radar" Middittee, operator, who was wearing a green eyeshade and earmuffs. He peered into his machine. The room was completely dark except for his lighted radar scope.

"What's up?" we asked innocently.

"You tell me," said Radar. "I'm getting mixed signals on this (expletive deleted) machine. According to my radar, it's snowing outside."

"That can't be," we advised. "This is the middle of July. It's 95 degrees."

"Oh, my God — that's what I was afraid of. This radar can't see straight. Is it raining?"

"Heavy rain," we said. Radar took off his earmuffs and put up his umbrella.

"When's the last time you were outside?" we wanted to know.

"I don't go outside. I live down here so I won't be affected by the weather. I love it. No weather in here at all so I'm not biased one way or the other. I'm completely impartial."

As we left, Radar pleaded, "Please call me once or twice a week and let me know what's going on outside." We promised to help.

We stopped by to see how the training program for new observers was coming along. The government has built an indoor weather laboratory to simulate outside conditions. It was raining at the time.

Inside the lab the student observers are subjected to wind, rain, snow, hail, and fog, just to get the feel of it. We wondered if they would ever be allowed outside again. We certainly hope so.

Wine, Women, or Song

> He was given the option of giving up wine, women, or song.
> He gave up singing.
> *Anonymous*

Sweden has run out of ethanol gasoline additive for its buses but has come up with an ingenious solution. The Swedes are instead now buying red wine from Spain as a substitute. The Swedes can always think of something.

In the past the Swedes drove on the left side of the road. When they voted to switch over to the right side, the government decided on a phased transition. They switched only the buses over to the right side for a week as a trial run. It did not work out very well. There are also some problems with wine in the fuel.

The wine has affected the buses. Drivers report that they can't drive in a straight line and that sometimes the buses make a wrong turn with no advance notice. Some buses even tried to switch back to the left side of the road. Mechanics have added Alka-Seltzer to the fuel tanks.

Several suspicious persons have been found lying under the buses while parked in the garage at night. After apprehension and questioning by police, these derelicts admitted they hoped for some excess drippings from the bottom of the fuel tanks. The police gave them extra Alka-Seltzer and released them after they promised not to bother the buses any more.

But the very next day in broad daylight several of the same miscreants hitched a ride under the buses while the buses were moving in traffic. The results were serious: several of these passengers disappeared in transit.

During lunch breaks bus drivers siphoned fuel out of their tanks to go along with their roast beef sandwiches. Red wine is ideal with red meat.

It is too early to tell whether this experiment will work out in view of the problems encountered. The French are now trying out both red and white wine in their buses; they needed white wine to go with their fish.

A problem arose in Paris. Instead of riding the bus the passengers insisted on running along behind to inhale the exhaust fumes. Wine experts rated the exhaust as having a superior "bouquet."

The public besieged one filling station demanding wine straight from the pumps. The pumps were labeled "red" and "white." Patrons brought their own 5-gallon cans.

Let's hear it for the Swedes and the French. They may be on to something environmental here.

We'll drink to that. *Skoal!*

The Dinosaurs Are Coming Back

During the next 30 million years the bird arrived, and the kangaroo, and by and by
the mastodon, and the giant sloth... and the old Silurian ass, and
some people thought that man was about due...Then at last the monkey came, and
everybody could see at a glance that man wasn't far off now...

Mark Twain on evolution

The movie *Jurassic Park* brings back fond memories of the dinosaurs who lived between 66 million and 250 million years ago. Paleontologist Jack Horner has dug up dozens of dinosaur skeletons in Montana. These particular dinosaurs never did get to evolve; they were buried by a volcano. Horner now believes that dinosaurs were warm-blooded, friendly, fun-loving creatures, not nasty and mean as they have been pictured.

Now some environmentalists would like to bring them back just for old time's sake. Believe it or not, that may now be possible. The geneticists have gotten into the act with theories about DNA and cells and genes and how to resurrect these ponderous beasts, among other species. This is revolution, not evolution.

We don't begin to understand how all this would work, but we think it goes something like the cloning you've been hearing about. You dig around and find some pieces of a dinosaur, maybe even an embryo in a dinosaur egg. You then scrape a nucleus from a cell and stick it into a host cell and implant it into a hostess, say a milk cow, we think.

Nature takes its course, and the cow delivers a dinosaur. You can imagine the startled look on the cow's face.

There is some evidence that the turkey has evolved from the dinosaur. We can't understand how that would be possible, although some dinosaurs did fly. This sure puts a different light on Thanksgiving. You may be forced to eat your turkey from a different perspective.

All of these genetic developments conjure up limitless possibilities. Maybe historians could scrape up some old cells from Christopher Columbus. We could implant these and bring him back in time for his next birthday. He would be a tremendous drawing card. Pity the poor cow.

Jimmy Carter, George Bush, and Harold Stassen could donate some cells and come back to run for office again in a million or two years when the right political opportunity presents itself. We would not want to be implanted in a turkey egg, however.

You may have read about people who are being freeze-dried at death in hopes of returning at a later date. This procedure avoids the undertaker, but the DNA route may be a preferable alternative. We also prefer it to being dehydrated and then rehydrated.

We are not sure whether this DNA process applies to inanimate objects or not. If it does, imagine the endless possibilities: We could resurrect the *USS Lexington* with some scrapings from Corpus Christi. Or try to imagine the consternation of the cow who gave rebirth to the San Carlos Hotel. We should be getting scrapings from the *USS Forrestal* right now before she is melted down into razor blades.

We can now plan much better for the future. Everyone should be on the lookout for scrapings from various persons or objects to be saved for future resurrection. What a great way to preserve history because we're not at all sure what monkeys will evolve into next.

Think of bringing back Harry Truman. We could certainly use him right now.

If we're ever reconstituted, we would prefer to come back as a dinosaur. A dinosaur might carry some weight with the government.

Unendangered Species

> I smell a rat, I see him floating in the air, but mark me,
> I shall nip him in the bud.
> *Sir Boyle Roche*
> *Member of Parliament, 1750*

Our endangered species are disappearing, but we have unendangered species which refuse to disappear.

In Louisiana we find the nutria, otherwise known as swamp rats. They resemble a muskrat and have a voracious appetite for flora and fauna. They are eating up all the vegetation in the Louisiana bayou country.

Believe it or not, a Louisiana gentleman imported only eleven of these junior monsters as pets. They soon multiplied to 200. Along came a hurricane, and all 200 got loose and scattered about in the swamps, there to multiply like rabbits. There are now 2,000,000 (2 million!) nutria. This is prolific.

What to do with these rodents? The Cajuns are experimenting with nutria chili. They throw in so much hot pepper that you can't tell you're eating swamp rats. This might be one way to eliminate these pests; eat them all if the hot sauce doesn't run out. Cajuns say they taste even better than snakes or alligators. This is comforting to know.

In Guam we find the brown tree snake. Authorities believe one pregnant tree snake sneaked into Guam; now there are 2,000,000 of these critters. They are eating all the birds on Guam.

Nobody is sure how to handle these snakes. Guam has called in some Cajun cooks and a supertanker full of hot pepper sauce to try and find a suitable recipe.

So far it is believed several of these pests have been sighted in Texas. Authorities fear the snakes will stow away on ships and airplanes and spread everywhere. They may be headed our way.

Shifting to Florida, we are now blessed with swamp eels. This slimy, air-breathing type fish can travel over land if necessary.

The eels have not yet attacked humans but are eating up all the fish in Florida. The eels are so slippery they are hard to catch. The Cajuns could try them in their chili if you can stand the thought. We don't yet know how many millions of these slithering creatures exist, but look out. We may need more hot pepper sauce.

Finally, alligators, once faced with extinction, now overrun Florida swamps and bayous. Alligators eat chickens, steak, dogs, small children, footballs, and other protein. Needless to say, now protected, they thrive by the millions.

One Florida alligator connoisseur recently died and left his 12,000 alligators to his son. The son was so emotionally attached to the beasts that he hated to liquidate his father's estate and lose his scaly friends.

There is a good market for alligators. Some people eat them. Asians love alligator hides. In this country, alligator purses, shoes, and golf bags sell for anywhere from $500 to $25,000. Think of that!

How can we endanger these species? We'll have to eat as many as we can as long as the hot sauce holds out. In the meantime, we should put them to good use. The nutria have been turned loose in the kudzu, but the kudzu is gaining on them. Alligators can eat bureaucrats if they don't choke on them. Right now we're losing to the snakes and eels.

Darwin would love this scenario. Survival of the fittest.

Heads Up

> It has recently been discovered that research causes cancer in rats.
>
> *Anonymous*

Science is outdoing itself these days. Cloning animals is all the rage. Gene splicing and all sorts of scientific hanky-panky fill the news. What's next?

Well, it's already here. Biologists have produced headless tadpoles which will grow into headless frogs. You will still be able to eat the legs, but science does not stop there.

Medical researchers now surmise that they can grow headless people. What on earth for? So they can "harvest" body parts which are in short supply for transplanting. This certainly is a worthy humanitarian goal.

We are trying to figure out other uses for all these people who will be running around without any heads until it's time to harvest their parts. We approached psychologist U.B. Nutz at his laboratory in Mosquito Flats, Florida. We wanted to find out the medical, sociological, and psychological impact of people with no heads. We found U.B. up in a tree contemplating people consisting only of body parts. "What's the scoop?" we asked.

"That's a tough one," he said. "I can see lots of problems."

"What kind of problems?"

"Well, suppose these guys get elected. They'd look funny sitting around in Congress with no heads."

"That might not be a bad idea. You could program their genes so they always vote right. They wouldn't pay any attention to lobbyists."

"Yeah, we can do that. No filibustering and all that nonsense."

"Wouldn't headless bodies eliminate a lot of mental problems? No more absent-minded professors. No forgetful shoppers who lose the shopping list, the car keys, or their cars in the parking lot. These guys will make ideal husbands."

"You're right. These guys are no-brainers. But I can see a lot of trouble with the barber shop union."

"How's that?"

"What would these people do for haircuts without any hair? The barbers will be up in arms."

"Couldn't you just grow some hair on top of their necks?"

"I guess so. Or else we could just stick a toupee on top and let it go at that. But the barbers will still be mad as hell."

"What else?"

"Great for education. No more over-crowding. These people wouldn't do worth a damn in school."

"How would they get around town?"

"Not on motorcycles. No place to put their helmets."

"Can you think of any other problems?"

"No, but this would solve the need for football linemen."

"What do you mean?"

"It's hard to find people big enough and dumb enough just to squat down in the middle of the line and jump out at people. These guys would be perfect. They wouldn't have to think. They wouldn't need helmets. We put the plays right in their genes; they couldn't forget 'em. No screw ups."

"What about salaries?"

"What salaries? These guys wouldn't need any money. They wouldn't know whether they were getting paid or not. They wouldn't argue. No more agents, no strikes. The owners would love it."

"By George, you've convinced us. This sounds like a great idea."

"Yeah, until some smart aleck scientist puts heads back on these people, and then we're right back where we started from."

A Wisconsin senator put it this way: "I'm in favor of letting the status quo stay as is." Nutz!

The Mad Scientists

> There are a number of things that increase sexual arousal, particularly
> among women. Chief among them is the Mercedes-Benz 380SL convertible.
> *P.J. O'Rourke*

Some recent scientific discovery is wild.

We learn that, scientifically speaking, certain smells arouse us, but it turns out they are certainly not the odors you would associate with romance. Now they are saying the correct smells will do the trick; you don't even need a Mercedes-Benz.

It seems hard to believe that women are turned on by smelling cucumbers, according to science. Then what smells turn women off? Scientists say cherry and after-shave lotion. Men have been dousing themselves with Old Spice, thinking to snare the lady of choice. Forget it. Rub on some cucumber; bring on romance.

What about smells for men? It appears they are aroused scientifically by smelling pumpkin pie. Pecan pie may work the same way. The ladies should forget Chanel No. 5 and dab a little pumpkin pie behind each ear. From now on don't eat cucumbers and pumpkin pie; they serve a higher purpose.

Bankers have called on scientists to stop bandits from ripping off customers at automatic tellers. The latest wrinkle is an auto teller which will read either your eyeball or your fingerprint for positive identification. Just look into the machine or mash your thumb up against the glass; the machine does the rest. If you aren't really you, no withdrawal. Scientists claim eyeballs are even safer than fingerprints.

Scientists think of everything. Suppose a bandit cut off your finger or took out your eyeball and presented it to the teller. Gotcha! The darn machine can tell whether you're attached or not. Detached parts won't work. We hope the criminals will figure this out before they cut off any of your identification.

Cloning is now a popular subject. If that weren't controversial enough, now comes some nutty professor who wants to crossbreed various species, including humans. Good Lord! Imagine crossing a rabbit with Bill Clinton. There could be some amazing results.

Cross Michael Jordan with a giraffe, and he could dunk a basketball 25 feet in the air. Breed a long-jumper with a kangaroo. Mate high-jumpers and pole vaulters with condors; cross 100 yard dashers with cheetahs.

Ye gods, no record would be safe. People would be jumping and vaulting hundreds of feet into the air; they would need parachutes to land. Track stars would run 60 m.p.h. Where would it end?

Crossing a human with a monkey would be a *BIG* mistake. Civilization would be going backward instead of forward, devolving instead of evolving. Devolution would take us back through the Garden of Eden to homo sapiens to Neanderthal to Pithecanthropus Erectus to apes to amoeba to zilch. Civilization would implode along with the universe in the year 60 billion. Nothing left, not even us monkeys.

Woe is us.

The Sky's the Limit

It's an impossible situation, but it has possibilities.
Samuel Goldwyn

According to Webster's Dictionary, levitation means "To rise or float in the air, esp. as a result of a supposed supernatural power that overcomes gravity."

Magicians like Copperfield or Blackstone or Houdini suspend people in midair all the time, presumably without visible means of support. But this is not supernatural; we all know there are some wires hidden there someplace.

Who would ever have thought that we could actually repeal Sir Isaac Newton's law of gravity? What goes up has to come down, according to Newton. But not any more.

Now scientists have done the magicians and Sir Isaac one better. They have actually levitated a frog. That's right — a frog, in midair. How? By using magnets with a magnetic force one million times stronger than gravity.

The frog floats around in air free and easy with no apparent ill effects. Why a frog? We don't know, but this maneuver brings up some interesting possibilities for mankind.

Think of all the things we could do if we could defy gravity and levitate. We could lift our cars up out of traffic jams and over-crowded parking lots. Airplanes would not need wings to fly. They would simply levitate to altitude where a simple set of propellers with rubber bands would propel them to the next airport.

Astronauts would no longer need rockets. They would levitate into orbit, thus saving a lot of fuel and rocket plumbing and plumbers. A lady with a big hat sits down in front of you in the theater; you can't see over her. You simply levitate twelve or eighteen inches and hang there with a perfect view. If the guy behind you can't levitate, he's in trouble.

There could be some other drawbacks. If we cut out too much gravity, we might lose the moon. Skinny people would be suspended in midair unless they gained some weight or carried barbells. The wind would blow them away unless they were anchored. Walking in air would be like treading water; you would have to carry a propellant to move around.

We thought of the effect of no-gravity on athletes. You can just imagine the possibilities. For instance, we thought about pole vaulters. What would happen to them? For some answers we went to track coach High Barr.

"Hi, High," we said.

"Hi yourself," said High.

"Have you heard about levitating frogs with magnets?" we asked.

126 Warren Briggs

"We sure have. We ain't interested in frogs, but we're excited about athletes, especially hurdlers, high jumpers, and pole vaulters."

"Tell us more."

"Well, right now we can only vault about twenty feet. We got to fight gravity all the way up. Just think what we could do with no gravity. The sky's the limit."

"How would it work?"

"Well, we just strap some big magnets on our vaulter. When he's ready to go, he flips the switch on. Wheee! World's record."

"How high could he go?"

"That's the only problem. He's got to turn the damned switch off quick enough; otherwise he could be airborne."

"Would that interfere with commercial airline traffic?"

"That's another problem. He would have to call in for landing instructions when he's in the traffic pattern."

"When he turns off the magnet, wouldn't there be a hard landing?"

"Oh, Lordy, yes. That's a big drop. Our guys would have to wear parachutes. Then there's the danger the switch would stick. God almighty, we might launch these guys into orbit. That would be awful. There ain't no gold medals for space walking."

So you can easily see there's good news and there's bad news when you fool around with gravity. The law of gravity was not made to be broken.

No Pain, No Gain

> I am dying with the help of too many physicians.
> *Alexander the Great*

A hospital is a miraculous place. We recently visited our friend Sherman Shipshape there.

He was strapped to his bed with hoses running in all directions from both ends; two tubes in his nose, suction, oxygen, catheter, glucose, drugs, you name it, Sherman had it all. The place looked like a plumbing supply company. All of the plumbing was strung up on a metal pole which was really a hat rack with wheels.

Sherman sat perched on his bed, looking majestic and triumphant like a king on his throne. He did not appear to be in any pain.

"I thought you were supposed to be hurting. Aren't you sick?" we asked him.

"No pain, no gain, as they say. Sure I'm hurting, but I'm gaining on 'em."

"How's that?"

"See this black box here with the blinking lights? That's chock full of morphine. Whenever I need a hit, I just press this little black button here. Wheeee! I'm high as a kite."

"How long does it last?"

"Oh, it's good till I have my martini."

"Good God, man, you can't have martinis in the hospital."

"Who says so? You see that bag of clear-looking stuff hanging there? That's my IV. They punch a hole in me and a hole in the bag. Then they run a hose up there, and it drips in me 24 hours a day."

"Good Lord! You mean to tell me that bag is full of martinis? I never heard of such a thing."

"Not exactly. But I have a friend who comes in to visit during cocktail hour. He squirts the martini into the bag with that other stuff. No olive, though, so the nurses can't tell the difference, but I can. I feel so good. Makes me forget my operation."

We were horrified. We started to ring for a nurse to report this serious violation.

"Don't bother," said Sherman, "they only come when you don't need 'em."

Then he climbed out of bed. We tried to stop him. "What are you trying to do?" we pleaded.

"I'm going out for a cigarette. They won't let me smoke in my room. I can't even sneak one, too many fire alarms. If I set one off, I have to pay the fire department."

"Where are you going?"

"I'm going outside to smoke with the rest of the emphysema patients."

"How can you walk with all those hoses hanging out?"

"Just watch me." He unplugged himself, hung all the hoses on his pole, and started off down the hall. He didn't seem the least bit embarrassed wheeling his urine along with him in a bag strapped to his pole in full view of God and everybody.

All the other patients got out of bed just to see Sherman on parade. He got lots of nice comments: "great production," "exquisite bouquet," "excellent color," "full body," "good volume," "satisfying ambience." It sounded like Chateau Lafite Rothschild 1949. These patients were obviously experienced urine watchers.

When he got back to his room, Sherman looked very satisfied with himself. "Did you hear all the nice things they said about my urine? They can tell I'm getting better; they're experts."

"What can you eat?" we asked.

"Nothing but ice cubes right now. I just had surgery."

"Minor surgery?" we asked.

"No, major surgery. Minor surgery is surgery somebody else is having." We were reminded of that old Spanish proverb: "There is no better surgeon than one with many scars."

"Aren't you hungry?" we asked.

"Sure. I'm dying for a cup of coffee. I'm trying to figure out a way to stick it in my IV."

"The next thing you know you'll be trying to smoke thru your IV."

"Hey, great idea! How would I put the nicotine in there?"

We had to leave before he got any more crazy ideas. Sherman was the most innovative patient we had ever seen, and the most sneaky.

All's well that ends well.

Never go to a doctor whose office plants have died.
Great Quotations, Inc.

Daily we receive all sorts of free medical advice through the mail along with ads for the pills needed to cure whatever ails us.

These unsolicited diagnoses cover every disease or symptom known to man or woman: diverticulosis, endometriosis, otorhinolaryngologosis, prostratitis, agrogooberacrophobia, and other horrible-sounding ailments.

We can't understand all these big words, let alone spell them. Doctors work in a strange world of words all their own. How do we find out what all these medical terms mean?

We recently received through the mail a brochure entitled "Medical Terminology for the Layman." We hope the following definitions will help you better understand your doctor:

ARTERY	The study of fine painting
BARIUM	What you do when C.P.R. fails
BENIGN	What you are after you be eight
CAESAREAN SECTION	A district in Rome
COLIC	A sheep dog
CONGENITAL	Friendly
DILATE	To live long
FESTER	Quicker
G.I. SERIES	Baseball games between teams of soldiers
HANGNAIL	A coat hook
MEDICAL STAFF	A doctor's cane
MINOR OPERATION	Coal digging
MORBID	A higher offer
NITRATE	Lower than the day rate
NODE	Was aware of
ORGANIC	Church musician
OUTPATIENT	A person who has fainted
POST OPERATIVE	A letter carrier
PROTEIN	In favor of young people
SECRETION	Hiding anything
SEROLOGY	Study of English Knighthood
TABLET	A small table
TUMOR	An extra pair
URINE	Opposite of you're out
VARICOSE VEINS	Veins which are very close together

With the help of this list you should now be able to converse more intelligently with your doctor. If he or she can't understand you, loan him or her a copy. There should be a copy in every doctor's waiting room.

Moooo

The best way to pass a cow on the road when cycling is to keep behind it.
R.J. Mecredy
Cycling Magazine

"A cow will produce more milk if she sleeps on a waterbed," according to a recent scientific study.

We never cease to be amazed at modern science. We have a bunch of mad scientists investigating every possible angle. They have even come up with medicines for which there are no known diseases.

Scientists are the same people who brought us kudzu to take care of erosion. Now kudzu is burying us alive.

Since we have no experience in dairy farming, we sought out Angus "Milk 'em Dry" Kowpize, an experienced dairyman with a herd of milk cows.

"Have you heard about the expert from Washington who recommends that cows sleep on waterbeds?" we asked Angus.

Angus threw his battered straw hat on the ground and stomped on it viciously.

"What was that tantrum all about?" we asked.

"That was all about these so-called experts who keep telling me how to run my business. I'll bet that guy never seen a cow. Do they have cows in Washington?"

"No cows, just bureaucrats who are there to help you. Why not listen?"

"I can do without their help. You ever tried a waterbed? My wife and I did, and the damn thing sprang a leak. Water all over the place. We almost drowned."

"We're sorry, but that was just an unfortunate accident. People all over the world sleep on waterbeds."

"That may be, but they'd better wear life preservers. I ain't dried out yet."

"How about just giving it a try? More milk means more profit."

"I buy 100 waterbeds, and there goes my profit already. Have you ever seen a cow try to lie down? It's awful. It ain't pretty. Can you imagine trying to get a cow up onto a waterbed?"

"It may take some practice."

"Practice, my foot. Supposin' all them beds busted? Good God, man, that's a flood. We'd have to man the lifeboats. We couldn't pump fast enough to save the barn. Might lose the whole herd. You can't put a life preserver on a cow; it don't fit."

"That's the worst case. You're a pessimist."

"A pessimist is an optimist with some experience so I ain't too optimistic. I've had enough of these experts. Next thing ya know they'll want me to read bedtime stories to them cows. Nuts to that. I ain't got time for this nonsense."

We left at least feeling we had tried to be helpful. Yves Saint Laurent bought a herd of buffalo because he said his "cows started to look tired." We'll go talk to him. Maybe he'll try the waterbeds.

ON SPACE

The end of the human race is that it will eventually die of civilization.
Ralph Waldo Emerson

People don't go there anymore. It's too crowded.
Yogi Berra

Gentlemen, include me out.
Samuel Goldwyn

Make plans. It wasn't raining when Noah built the Ark.
Unknown

Mars, Dead or Alive

Mars is essentially in the same orbit (as earth). Mars is somewhat the same distance from
the sun, which is very important. We have seen pictures where there are canals, we
believe, and water. If there is water, that means there is oxygen.
If oxygen, that means we can breathe.
Vice-President Dan Quayle

The rock from Mars still has everybody excited, especially at UFO headquarters in
Gulf Breeze. Scientists are still looking through their microscopes at what may be bugs
or some other form of life in the rock. Scientists, like economists, don't agree on anything.

If there was life, then how did it get to Mars? Maybe God started off the universe
with a big bang and then let nature take its course. Or maybe he did a complete package deal right off the bat with a Garden of Eden, Adam and Eve's brother-and-sister
Dick and Jane, a serpent, an apple, etc. Hopefully the bug in the rock did not come
from the apple; otherwise Mars might have started out rotten instead of evolving as we
have.

We don't pretend to know the real answers to all of this. We just hope for the sake
of the UFO folks that they do find something alive out there. How else could UFO's
get to Gulf Breeze?

Whatever the case, we certainly need to pursue the Mars story further. We just
hope and pray this famous rock really did come from Mars. Some experts have questioned that fact. It would be a terrible shame to discover later that the rock just fell off
the back of somebody's dump truck. That could just about demoralize the scientific believers, the UFOers, and whoever it is who is up there on Mars.

Just recently, Dr. Shannon Lucid, Ph.D., returned to earth after spending *six
months* in the Russian Mir space station. Can you imagine six months floating around in
midair and talking to a bunch of Russians? The previous American Ph.D., who spent a
considerably shorter time on Mir, almost went nuts. He said he was bored to death.

The U.S. is already planning to fly to Mars to explore first-hand. The UFOers can hardly wait. Good Lord! Think of it! Mars is 475 million miles from earth. Earth and Mars both orbit the sun. We have already sent out unmanned space ships to investigate Mars.

More such space vehicles will eventually land on Mars and send out miniature Cadillacs to snoop around and report back. They may even pick up more rocks and send them back by Federal Express. The first landing will take place at 2:00 o'clock Central Mars Time (CMT) on the 4th of July, 1997. More unmanned vehicles will follow.

Eventually a manned or womaned rocket ship will blast off to Mars to land and explore personally to see what is really up there. The voyage will take *one year. A YEAR!* Try to imagine spending a year cooped up in a large, pointed, tin can-type vehicle waiting to get to Mars to find who-knows-what. Then a year to come home. What would you do on the way to keep your sanity?

Just for fun, think of yourself on a week's vacation locked up in a Volkswagen bug and watching paint dry and grass grow. Then multiply that week by 52 to get to Mars.

You had better have your head screwed on tight! Otherwise they had better have a nut-house on Mars.

Rocks, Bugs, and Worms

> Space travel is utter bilge.
> *Sir Richard Wooley, 1950*
> *Royal Astronomer of Britain,*
> *on the likelihood of space exploration*

Mars is a hot topic these days.

It seems that in 1986 somebody exploring the South Pole stubbed his toe on a bright green rock. Scientists brought it home and ground up pieces of it so that they could examine it under a microscope.

Zounds! The scientists saw something that looked like fossil remains, maybe a microbe or a bacteria or an amoeba or something. It was a long, skinny-looking thing. Mike Royko in Chicago thought it looked more like a worm to him. Or maybe it was a bug. Whatever it is, the scientific community is drooling along with UFO observers who need to find life out there someplace, Mars in particular. They need to locate somebody who can drive the UFO's they keep seeing.

4.6 billion years ago or so, two big clumps of gas and dust coagulated. One turned out to be earth, the other Mars. Some time later a meteorite bumped into Mars and knocked off chunks, including this rock. The rock floated around in space for a while until earth's gravity finally grabbed it and pulled it in.

To get some perspective on the rock and the bugs or worms, or whatever they were, we interviewed I. Seenem, a UFO expert at UFO headquarters in Gulf Breeze. He was also drooling over the prospects.

"I tried to tell you there was life out there," he crowed. "But, no, you guys knew it all. Now maybe all you know-it-alls will wake up and get with the program."

"This isn't exactly an alien space ship we're talking about here," we reminded him. "We don't even know for sure whether it's a dead bug of some kind. And if it is, when you UFO nuts finally do get to Mars, how are you going to talk to an amoeba?"

"There you go again. That bug is billions of years old. That bug could eventually have evolved into an alien who could operate a UFO. Obviously an amoeba couldn't do it."

"But some folks still don't believe in evolution," we reminded him. "They believe God cranked up the universe and everything in it. It doesn't say anything about this rock in the Bible."

"Well, this all happened a long time before the Bible. After all, God created Adam and Eve here on earth. Maybe he did Jack and Jane on Mars."

"That theory won't work," we said. "Adam and Eve only wore fig leaves, and it's freezing cold on Mars. Below zero."

"You're not thinking straight, as usual. Mars was originally warm; it froze up later. Besides, if Jack and Jane needed earmuffs and fur coats, God could certainly have handled that without any trouble. After all, he came up with fig leaves when he needed them."

We came away wondering about all of I. Seenem's theories. We never have bought all his UFO stories, and there is just too much controversy about creation versus evolution.

This subject needs further examination.

Star Light, Star Bright

> Maybe this world is another planet's hell.
> *Aldous Huxley*

The pictures coming from the Hubble Space Telescope are stupefying. Astronomers are finding all sorts of new neighborhoods out there in the universe.

Light travels at *186,000 miles per second*! That means that some of the Hubble pictures we are getting now may be *10 billion years old* or older. Astronomers estimate the age of the universe at between 8 billion and 15 billion years. Such numbers are incomprehensible.

We recently questioned astronomer Ora Bory Allis about these astounding figures and findings.

"How did the universe get there?" we asked.

"Darned if I know," said Allis. "But there's plenty of it out there. We keep finding more chunks."

"Are there any people out there?"

"You're darned right there are! I track them every day. I'm the only one who knows about them. Nobody else believes me."

"How do you know they're there?"

"Ever see a star blink? That's a secret code. They sent me the key to the code so I can decipher their messages," he explained.

"What do they say?"

"They say stay the hell away from their planets."

"Why so unfriendly?"

"They don't want to be bothered with foreign aid or peace-keeping missions or welfare or any of the rest of our charity. Who needs it?"

"They must be Republicans," we said.

"No way. They're independents. They like Ross Perot, but they don't like what they see down here: AIDS, divorce, shooting, stabbing, Watergate, Whitewater, drugs, debt, gridlock, war, hurricanes, earthquakes—you name it, we've got it."

"Well, what's the news from up there?"

"They don't say too much. Remember, the signals are billions of years old. They don't talk to the press."

"Will they ever come to visit us?"

"Naw, it takes too much gas to go 10 billion light years. Besides, we might not be here when they get here."

"They may think more like us by this time."

"God forbid! I certainly hope not. Most of them hid in dust clouds to keep us from corrupting them. They say no news from us is good news."

"Think what they're missing."

"Yeah, I do: political campaigns, illegitimacy, pornography, TV, XXX rated movies, cigarettes, whiskey, condom ads, all that good stuff."

"We do have a few good things happening."

"Not much. They believe our press."

We had to agree with Allis. Those folks are better off up there by themselves. Lord knows where we'll be in another billion years.

Allis let us look through his telescope. The view of the universe is spectacular, astounding. How did the universe happen?

Then we suddenly remembered. One good thing for sure: God is out there somewhere.

Thank Heavens!

Mars on the Rocks

> Changing your mind does not involve a transplant.
> *Robert Half*

We can't get Mars off our mind. Is it dead, or is it alive with bugs and aliens or something else?

We certainly didn't intend to get caught in the middle of the argument over creation versus evolution. That's like arguing about flat taxes or abortion; you can't win.

Twain goes for evolution; the Bible goes for creation. However life got started out there, the UFO folks have got to find it. Somebody has to fly those UFO's to Gulf Breeze. UFO believers are now pinning their hopes on Mars because of some fossil evidence of bugs found in a rock which fell off Mars and landed at the South Pole.

We previously reported some supporting evidence. Astronomer Ora Bory Allis claims he gets signals from aliens in space through the Hubble telescope which is floating around out there. The stars blink at him in code and that's how he gets his messages. He says the aliens don't want us messing around out there.

Aristotle McGoober in Mosquito Flat communicates with the deceased. He says if necessary he can also talk to dead aliens, but nobody has asked him. He doesn't know if he can talk to dead bugs.

Commodore I. Seenem, commander of the UFO base in Gulf Breeze, has actually seen aliens there. Generally sightings occur during cocktail parties. Seenem is chomping at the bit to climb into a rocket and blast off to Mars to meet whoever is up there. Actually this is the only way we will ever find out who is right and who is nuts and who is out there.

Scientists now say that Mars is so cold that life as we know it is impossible there, and there is no oxygen. But those facts don't daunt the true believers.

They think that there is hot water under the ice and that the inhabitants live in caves. They don't know what the aliens do for oxygen, but maybe the aliens inhale something else up there to keep going. These are the same folks who fly the UFO's.

Other believers take a more pessimistic view. They think that the people on Mars lived in steam-heated igloos and wore oxygen masks. Eventually the oxygen ran out, and the igloos overheated and melted. All that was left was snow and ice and nobody to fly to Gulf Breeze. This is the same frozen wasteland which the U.S. now wants to explore.

Frankly all of this sounds silly to us. We don't believe any of it; we have our own ideas.

We say whether by creation or evolution, we can't say which and don't dare to, there was life on Mars before the place froze up. There was plenty of fresh air, and there were Democrats and Republicans. They fought over politics just the way we do now.

Finally the Republicans took over their Congress and tried to cut spending. The Democrats revolted and manned the barricades. Civil war broke out. Since there were no guns in those days, they threw rocks at each other. One of those rocks fell off of Mars, the same rock with the bugs in it which the scientists found at the South Pole.

That was the bloody end of everything on Mars. Life disappeared along with the contract on Mars, the balanced budget, term limits, welfare as we know it, line-item veto — everything. All that's left are the rocks with the bugs in them.

There but for the grace of God go we.

Up, Up, and Away

> It's always darkest before it's totally black.
> *Mao*

Gulf Breeze UFO headquarters is celebrating. There is dancing in the streets.

Why? Gulf Breeze has finally received confirmation that there is really a UFO out there and it's coming. The mayor wants it to land here and has wired Hale-Bopp offering no landing fees and free hangar parking.

How do we know for sure that it's out there? Thanks go to the Heaven's Gate cult which recently sacrificed itself to outer space.

The news is full of Heaven's Gate, previously known as Higher Source, Total Overcomer's Anonymous, and the Next Level Crew. Crew members deserted their families and camped out together in black tennis shoes and short haircuts, ate pizza and drank soda pop, ran computers, built a house out of used tires, then rented a mansion in Rancho Santa Fe, California. Sam Kouchesfahani, who owned the mansion, was not pleased when he found out he was housing an army of fanatics.

The head fanatics were Marshall Herff Applewhite and Bonnie Lee Trusdale, otherwise known at various stages as Winnie & Pooh, Bo and Peep, Chip and Dale, and lately Do and Ti. She was a nurse, he a bisexual music professor with a checkered past. They met in the insane asylum.

The cult pretty much invented its own religion based largely on flying saucers. Somehow a gigantic flying saucer was scheduled to come down and pick up their souls and take them to a higher level somewhere in the promised land.

Heaven's Gate was a very platonic outfit as the members practiced celibacy. Do and other males even castrated themselves to make sure there was no hanky-panky. They considered their bodies "vehicles." To wash their vehicles they drank "master cleaner," a concoction of lemonade, cayenne pepper, and maple syrup.

Along came the Hale-Bopp comet which showed up for the first time in 4000 years. The Pharaohs were the last people to see it. Voilà! This was the heavenly sign they were all waiting for.

Some amateur nut in cyberspace claimed he clearly saw on his telescope a big UFO parked behind Hale-Bopp just waiting to swoop down and scoop up Heaven's Gate. This was too good to be true.

So, very unlike its usual conservative practices, the cult held its last supper in a gambling casino, played the slot machines, ate pizza, drank coke, and went to science-fiction movies. Then they ate applesauce spiked with paregoric, chased it with vodka, put their heads in garbage bags, and lay down to die for the cause, whatever it was. They died with their sneakers on. Their souls are scheduled to board the UFO, leaving their "vehicles" behind, much to Sam Kouchesfahani's displeasure.

Sad as all this may be, the UFO sighting has overjoyed the Gulf Breeze UFO Association which has also been waiting for a sure sign. Hale-Bopp is it.

One little Gulf Breeze lady was ecstatic. "I'm just dying to catch a ride on a UFO," she said. We think she should rephrase this slightly as it upset the mayor who has pleaded with the local group not to abandon its "vehicles". But our folks are excited waiting to be picked up, hopefully still wearing their bodies.

We asked astronomer Telly Scope if he could confirm the sighting of the space craft behind Hale-Bopp.

"Naw, I'm sorry, I thought I had something. But when I took another look, I seen it was a bug landed on my lens."

After all their sacrifice, we certainly hope Heaven's Gate didn't miss the boat.

Here They Come, Ready or Not

> You can observe a lot by watching.
> *Yogi Berra*

Paul Revere shouted, "The British are coming." If he were riding today, he would holler, "The asteroids are coming."

Asteroids, comets, and meteors are big hunks of rock, gas balls, globs of gunk, or whatever else we don't know, out there wandering around in space. Every once in a while one of these hunks hits us, slam-bam! They come in all sizes from golf balls to something as big as a mountain. Asteroids can measure 480 miles in diameter. That's a big chunk.

In a recent column George Will outlined the terrible consequences of a collision with earth. Little ones hit us every hour or two. A house-sized rock came within 65,000 miles in 1994. That's not very far in space numbers, too close for comfort according to astronomers. (The earth orbits the sun at 66,500 m.p.h.)

George Will continues: "Every once in a million years ... earth collides with an object a kilometer in diameter—large enough to destroy India.

"Last November only about 10 hours may have spared Manila or Bangkok from the impact of the meteor that instead dug a crater 176 feet wide and ignited acres of coffee plants in Honduras.

"... 65 million years ago a comet or asteroid perhaps 10 KM wide hit near Yucatan, creating enough environmental havoc to kill the dinosaurs and nine-tenths of all other species on earth.

"If one hit the Atlantic near Bermuda, it would generate a wave that would be 600 ft. high when it hit Manhattan at 500 m.p.h. and would submerge low-lying regions from Dublin to Hong Kong.

"Soot from fires started by the exploding comet might blot out the sun for a year, and acid rain would destroy much of the remaining plant and animal life."

My God, people, do you realize what we're talking about here? This is serious business. Somebody had better start doing something about these asteroids. We had better get ready for the attacks. We think there are people out there throwing rocks, big ones, at us. So watch out. We have to keep an eye on these monsters.

We called the White House to get a response. Nothing. They are too busy with Whitewater, Filegate, foreign contributions, and school lunches. "We can't be bothered with rocks," said a White House public affairs person down somewhere in the basement where he wouldn't even be able to see one coming.

Well, we can. As General Patton always used to say, "Do something, dammit, even it's wrong." Frankly, we're in a panic so we assembled a panel of experts to evaluate our chances of survival:

> General "Bomb" Bastick, USAF (retired), Star Wars, missile, and asteroid adviser.
> Aristotle McGoober, a psychic with offices in Mosquito Flat, Florida.
> Ora Bory Allis, astronomer stationed on top of the Sun Trust bank building.
> I. Seenem, UFO observer in Gulf Breeze, Florida.

We can now report some excerpts of testimony from the panel. (It was not called a commission because commissions never solve anything.) Consider the following statements from the experts:

I. Seenem - "I told you there were people out there. Nobody believed me. Now you know. I seen 'em myself. They're throwing rocks at us.

"We've got to get out there and sign a peace treaty and get them to stop. Secretary of State Madeline Albright is scheduled to go back on the next UFO. She's tough. She'll tell 'em like it is."

Ora Bory Allis - "It's been so foggy we can't see anything through our telescopes, but we've heard them moving around out there. They're up to something. Soon as the fog lifts, maybe we'll be able to spot something. But they're sneaky."

Aristotle McGoober - "I ain't no astronaut; I'm just a poor psychic. I can't tune them aliens in. I tried to contact Eleanor Roosevelt; she has connections out there, but she was talking to Hillary.

"I did get hold of Paul Revere. He said he's available if we need him. I don't know what else I kin do. But I do smell something fishy going on out there. I feel it in my bones."

General "Bomb" Bastick - "You're damn right they're out there. We're on full alert. Damn the torpedoes. We'll be ready if they throw the big stuff at us."

The panel has a plan. It calls for asteroid watchers stationed around the world with binoculars, just like the air-raid wardens in World War II who spotted enemy planes. When they see one of these things coming our way, they will report by walkie-talkie to General Bastick who will send the alarm world-wide by Internet. We hope the Internet will not be clogged up with students and Vice-President Al Gore. We won't have much time to spare.

Horses and riders will be standing by, just in case, to carry the alarm to every "Middlesex, village, and farm." Lanterns in the World Trade Center will signal the horseback riders, "One if by land, two if by sea." This way we're completely covered.

General Bastick will launch missiles and aircraft to shoot down the intruding rock or rocks. The general has also deployed rocket thrusters on all sides of the globe. Should his missiles miss the rock, he will fire the rockets on the appropriate side of the earth to move us far enough over so that the rock will miss and go right on by to hit something else.

We're still nervous, but Al Gore thinks this plan will work. It's worth a try. We can't just sit here and do nothing and wait to balance the budget in 2002.

Mars or Busted?

> Optimism is a little boy looking in a pile of manure and hoping to find a pony.
>
> *Unknown*

The Pathfinder mission finally hit Mars after 309,000,000 miles for $260,000,000 plus, or about 85¢ a mile.

The rocket hit speeds of 16,600 m.p.h. and then in just 4 minutes slowed to 140 m.p.h. and dumped its lander out by parachute.

The landing package, protected by huge inflated air bags, hit the ground at 23 m.p.h. and bounced five stories high. Then it bounced two more times until it finally settled down and stopped right-side-up after seven months in space.

Later a little six-wheeled robotic, Cadillac-priced Landrover rolled down a ramp and strolled around Mars sniffing rocks.

After several more of these trial balloons, NASA actually wants to send people to Mars. Some nuts have already volunteered to go. The round trip will take two years. Why anybody wants to go to Mars is beyond us. It gets down to -64° F there although it does warm up to -10° F in the sunshine. Balmy! There's nothing up there but dirt and rocks. The only advantage is that there is no government there ... yet.

We all wonder what benefits we'll get from these excursions. Not too many of us yearn to vacation in this Godforsaken place, but there's always some guy who wants to get away from his wife for a while.

What then do we reap from our hundreds of millions of tax dollars on Mars? For some answers we contacted local astronomer, geologist, and tattoo artist Rocky Stones in Mosquito Flat.

"Can you tell us what good these trips are doing for the average citizen?" we asked.

"You sound just like the rest of the pessimists. Why don't you lighten up and get a positive attitude? Be optimistic for a change."

"Well, we would if we didn't worry about what good our tax dollars are doing 119,000,000 miles from home. We're not interested in rocks."

"Let me give you some good examples," said Stones. "Did you see those big air bags that looked like balloons on the lander, and did you see how high it bounced before it landed safe and sound, not a scratch?"

"Sure, it was all over T.V."

"Well, there's your answer. How many times have you been on an airline and seen your bags scrambled, dropped, pitched, stomped, and generally kicked around? In Denver they even have an automatic baggage handler that automatically throws your baggage at speeds up to 50 m.p.h."

"So?"

"So you hook air bags to your suitcase. No matter what the airline does to your stuff, they can't kill it. It just bounces around safe and sound, no damage."

"What if they lose my suitcase?"

"They can't very well hide all those air bags."

"That sounds practical. Maybe Mars does have some possibilities. Tell us more."

"Ever had a package delivered that looks like it went through a rock crusher or got run over by a bus? Same deal. Tie on your air bags, and you're safe. The delivery man or mover can't dent it. He can't even get his hands on it."

"Won't this make delivery persons mad?"

"Sure. One postman got so mad he went berserk and dropped all his air-bagged packages off a cliff. Everything came through in perfect shape. He committed suicide; he couldn't stand it."

"I'm coming around. Any more good ideas from Mars?"

"They're endless. Remember when the governor tried to stop the Seminole Indians from bungee jumping? Well, bungees are out, air bags are in. Bungee jumpers don't land; they just hope they don't hit bottom. They end up in midair."

"But air bags hit the ground."

"Exactly, but they bounce. That's the beauty of it. Now the Seminoles can drop you off the Empire State Building, no sweat. You bounce around and land on your feet safe and sound if everything goes OK. Whatta deal!" Stones was ecstatic.

"Well, doggone, you've convinced us. Now tell us what to do with the rocks."

Spaced Out

> A lawyer is a learned gentleman who rescues your estate
> from your enemies and keeps it himself.
> *Lord Henry Brougham*

NASA, the Mir space station, and even the Mars Rover have had enough problems without piling on any more. Now comes still another glitch.

Three Yemeni (these are people from Yemen, wherever that is) are suing the Mars Rover for trespassing. *TRESPASSING*! Believe it! Can you imagine suing somebody 119,000,000 miles away?

The gentlemen from Yemen claim that their ancestors willed Mars to them 3000 years ago. They claim their relatives posted *No Trespassing* signs on Mars at the time. So what happens now?

The Rover's batteries will run down soon and won't be recharged. The Rover is pooped out from visiting rocks. In his weakened condition he recently stubbed his toe on a boulder. Before he runs out of juice and expires, Rover has hired a lawyer to defend him against the Yemeni. Rover is mad as hell, and he won't take this lying down or however a robot passes away.

We talked to attorney Abe E.S. Corpis of the law firm of Fetchum, Ketchum, and Soakum about the legal ramifications of this situation.

"Can Rover afford to hire a lawyer?" we asked.

"Oh, sure. He can't afford not to. We're already on the case. We'll counter-sue and soak the Yemeni."

"How can you handle a case 119,000,000 miles away?"

"Well, that is a small problem. We'll have to send Soakum up there. We're fitting him with a rocket and parachute right now."

"What if Soakum overshoots and can't find Rover?"

"That's another small problem. Rover can't travel too far. He's already exhausted, and he's only traveled 75 feet. Soakum will have to find him. He's taking his bicycle with him."

"What about the *No Trespassing* signs? Isn't that a violation of Yemen's property rights?"

"What *No Trespassing* signs? Our client denies everything. He hasn't seen anything but rocks."

"Suppose the Yemeni have witnesses."

"What witnesses? 3,000 years old? That's a laugh. What are they gonna do? Dig 'em up? My guess is they won't talk."

"Suppose Soakum lands beside a *No Trespassing* sign? Then what?"

"Don't worry. He won't see anything."

"What if there are still Yemeni up there walking around?"

"Well, that could be a problem. Soakum will be armed if he has to fight them off. Otherwise, he'll just take their depositions. They'd better cooperate or else."

"What if they get to Rover before Soakum?"

"Rover is programmed for that. He talks to nobody without a lawyer. Besides, he's too tired to talk. Soakum will do the talking when he gets there."

"Suppose the Yemeni don't speak English, then what?"

"They'll have a lawyer, and all lawyers speak Latin. Quid pro quo. Gaudeamus igitur. Cogito ergo sum. Non sequiter. Res judicata. The corpus is delicti. We habeas the corpus. E pluribus unum. Et cetera. They'll understand that."

"Will you go to trial up there?"

"No way. We'll get a change of venue and bring the case back to L.A. and Judge Ito. The plaintiffs won't have a chance. We'll play the robot card. Robots are a minority. We'll have robots on the jury."

"But all that won't bring Rover back to life."

"May he rest in peace after he pays our fee. We get 33% plus travel to Mars."

"It sound as if you and Soakum have everything under control. Do you have any last words for Rover and our readers?"

"Non illigitimi carborundum."

It's a Bird

> Serendipity is looking for a needle in a haystack
> and finding the farmer's daughter.
> *Unknown*

Serendipity is also finding a UFO.

UFO Headquarters in Gulf Breeze is always on the lookout for UFO sightings. Now the believers have found another one to investigate.

Reports from the vicinity of Eglin Air Base tell of a bright celestial object landing in that area. First reports indicated the object crashed, but UFO's don't crash.

UFO's land and emit little people who invade the territory and do God knows what. Sometimes they kidnap natives and take them for a ride. Sometimes they run for office.

Naturally, Gulf Breeze headquarters was ecstatic. Search parties fanned out to locate the landing site and verify the reports which came from reliable sources.

We always like to confirm these UFO's ourselves. The seat of knowledge in UFO matters resides in the U.S. Air Force so we contacted Colonel Hiram "Hi" Flyover who is the Air Force person in charge of UFO's and AWOL's. We found the good colonel looking through his spyglass for any sign of suspicious movement.

"Any incoming?" we asked him.

"Naw, just some pigeons and the Goodyear blimp. Somebody'll be reporting that damned thing so I have to be ready to investigate."

"You don't really believe in UFO's, do you?"

"Hell, no, it's just a lot of liberal propaganda. I've got enough to do finding AWOL's without chasing a lot of hot air. I see one AWOL coming this way right now," he said while peering through his spyglass.

"We hate to tell you, but we've got another sighting near Gulf Breeze and Eglin."

"Don't tell me that. What happened? A Hale-Bopp cocktail party?"

"No cocktails. We had four legitimate reports from around Eglin."

"Cocktails or no cocktails, that's just more hot air. Go talk to our Department of Denials. I can't take any more of this stuff." He threw his spyglass out the window.

We found the Denial Department down in the deep dark basement. Dr. N.O. Chance was running his mimeograph machine 100 miles per hour. When he saw us, he slammed on the brakes.

"How can you see a UFO way down here underground?" we asked.

"That's not my job. I don't want to see 'em. My job is not to see 'em."

"But we've just had another sighting down near Gulf Breeze headquarters."

"Oh, no, not Gulf Breeze again. That's supposed to be a dry county. Have those people been drinking?"

"No way. Actually this UFO was over near Eglin. That's a wet county, but we're just trying to confirm it."

"Well, I can't confirm it down here. What did the reports say?"

"First it was a bird, then a plane, then Superman, finally a UFO."

"What did Eglin see?"

"Nothing. They couldn't find a trace."

"Just as I expected. Stand back while I crank up my mimeograph machine and make my report. I need 200 copies."

"What will it say?"

"UFO unconfirmed. Report denied. It was only Superman."

Flying Rocks

> I don't see, Mr. Speaker, why we should put ourselves out of the way to serve posterity. What has posterity ever done for us?
> *Sir Boyle Roche, British Parliament*

A large rock, one mile wide, will pass earth on Thursday, October 26, 2028, at 11:07 a.m., according to the government. This rock is called asteroid 1997XF11.

First reports had this rock 30,000 miles from earth, pretty close for a big rock. A later report says no, it will be 600,000 miles, a slight discrepancy but close enough for government work.

In any event, not a very encouraging report. Even if this one misses us, there are hundreds more out there in space. If one hits us, what will happen?

Naturally Al Gore worries about the environment. The force of a hit can equal 100 atom bombs. For instance, if this rock splashes down in the south Atlantic, it will wash Florida away. Not a very happy thought.

Asteroid reports stirred little reaction from Congress. Florida Congressperson Rocky Smith, chairman of the sub-committee on flying rocks, couldn't be bothered.

"Hell, what do you expect me to do? We can't be worrying about stuff 30 years away. I'm interested in pork, not rocks. Go talk to NASA. That's their department."

NASA representative Rocket Jones was not overly concerned. "There's lots of junk flying around out there. Our only job is to track it. Go see the Defense Department. That's their business."

The Secretary of Defense perked up his ears. "You don't say," he said. "I'll take that up with the Joint Chiefs of Staff." They were too busy.

We finally found a Navy person who would listen. "Have you heard about the asteroid?" we asked him. He had. He laid it on the line.

"Remember when the battleship Missouri got stuck on a sand bar? Well, the Air Force offered to hook a B-52 onto it and pull it off. That made the Navy mad. We just told the Air Force to take their hot air and go out and blow it off."

"That wasn't a very nice thing to say."

"We meant it. We hate the Air Force. This asteroid is their problem. The Navy isn't involved unless the damned thing hits the ocean. Then we're in business; we're in charge of water."

We were running out of possibilities, but we visited Colonel Hi Flyover, Air Force person in charge of flying objects, including rocks. "Are you guys prepared for 1997XF11?" we wanted to know.

"What the hell is that?"

"That's the asteroid."

"Oh, that. We're out of it. We only go up to 100,000 feet. We can't shoot down something 30,000 or 600,000 miles away. How far is it, really? Congress will have to give us some more money. We can't afford to shoot at flying rocks on our measly budget."

In desperation we called on Herman Hobbgood, former sawmill engineer in Mosquito Flat, Florida. "Frank, aren't you worried about this flying rock?"

"No way. I know what to do if it gets too close."

"Thank God somebody knows," we said, greatly relieved. "What's the answer?"

"Simple. It don't take much to move an asteroid. I figure we stick an ordinary Sears leaf-blower on one side of the rock, turn 'er on, and move that thing out of the way. Won't cost much to do it."

"Sounds too easy to us. What if that doesn't work?"

"We go to plan B. Somebody shouts 'rock incoming', and we all rush over to one side of the world and blow on it. That'll do it if the world don't tip over."

"But what if that doesn't work?"

"Duck."

ON SPORTS

The thing I like about baseball is that it's 90% mental. The other 50% is physical.
Yogi Berra

I made more than President Hoover, but I had a better year.
Babe Ruth

One way to stop a runaway horse is to bet on him.
Anonymous

My golfing ambition is to live to be 100. I always wanted to shoot my age.
Unknown

Professional football is rough. The linebackers take hostages.
Anonymous

A difference of opinion is what makes auto racing.
Mark Twain

Football is not a contact sport, it is a collision sport. Dancing is a contact sport.
Vince Lombardi

Jocks

If you don't throw it, they can't hit it.
Lefty Gomez

Sports are not very sporting anymore.

The new spirit is to *WIN* at any cost, and sports are now very expensive. It's kill, kill, kill, and money, money, money. Lately there has been a rash of serious incidents.

In Texas a weird mother plotted to kill a high school girl who was her daughter's rival to be head cheerleader. This was strictly amateur night compared to the professionals.

Next a German nut stabbed tennis player Monica Seles in the back literally so that German Steffe Graf could become number one in women's tennis.

Now, ladies figure skating champ Tonya Harding's part-time husband and her idiot bodyguard are accused of hiring thugs to bash in Nancy Kerrigan's knees so she couldn't skate and win. Lovely.

Soccer fans in Europe riot and trample each other to death. Little League parents punch each other over a kid's game. Hockey players commit mayhem in what looks like a refereed riot.

What is the sports world coming to? Consider pro football, where they're out to mangle quarterbacks who get combat pay for hazardous duty.

It's true that pro football players are not supposed to be out there playing drop-the-handkerchief for fun, but the pay for these pigskin gladiators is obscene: $400,000 average, $10,000,000 for a few. You can buy a lot of good schoolteachers for $10 million.

College football has succumbed to the big money game and turned "pro." The college players normally don't get paid directly, just quietly by the zealous alumni out to *WIN*. The players do get lots of perks and big bonuses to go pro. They don't even wait to graduate before getting drafted by the NFL. Academics are incidental.

Bobby Bowden and Charlie Ward at FSU want players to be paid for the proposed college bowl playoffs to find out who is number one in college football. Who gives a damn? Is it really that important and critical to know? Eugene McCarthy once said that "...coaches have to be smart enough to understand the game and dumb enough to think it's important."

Miami with nine wins and three losses thinks the end of the world has arrived. College coaches are fired for 8-3 records. Coaches break the rules, and their teams are punished with probation. No TV, no bowls. Fatal!

Education is not all that important. Just *WIN*! There is the famous story about the football jock who had to take a special math exam to be eligible.

The question: "How much are 2 plus 2?"

His answer: "Four."

His coach was horrified. "Oh, my God! For Lord's sake, please give him another chance."

Sports are big business, but they are too big, too much money, too much gambling.

We think the ancient Romans had the right idea: they had a low-cost sports program in the Colosseum. They watched the gladiators hack each other to death, and they didn't have to pay gladiators anything like running backs. In fact, they weren't paid at all, just fattened up at the training table to make a colorful mess at the kill.

Then there were the Christians. The Romans didn't pay them anything either; they just turned them loose in the arena with the lions. Somebody forgot to feed the lions so they ate the Christians. Today FSU linebackers eat opposing backfields alive. Great fun, great sport. The Romans loved it all and bought lots of season tickets, just like FSU fans.

The Mexicans have the right idea. We think bullfighting is the answer to big money and overemphasis in sports. The U.S. can import this sport from Mexico without any tariff under NAFTA. This is another low-overhead sport. Instead of eleven high-priced men you have only a matador, a few minimum-wage picadors, and a lady who cleans up after the bulls.

You pay the matador pretty well but nothing like a quarterback. No scholarships. Matadors don't have to go to college. Said Professor Jose Acadeemus of the University of Mexico, "Academics are not required. You can't educate somebody who is dumb enough to fight a bull." Matadors are expendable and so are the bulls.

Bulls are cheap, just water and hay. Bulls are considered throwaways. Bullfighters are covered by their own household insurance policies so the whole operation is almost entirely profit. Jerry Jones, owner of the Dallas Cowboys, has just heard about this and is looking into it, but Emmitt Smith is too smart to fight bulls.

Why have "sports" become so important and so overemphasized? For lots of money and lots of blood. They are a tribute to our culture and our mentality.

Cut the Fat

> Don't worry about losing weight. You'll find it exactly where you lost it.
> *Robert Half*

We all have to go sometime. The question is how and when. There are several ways. Some say, "Thin is in. Fat is out."

At least that's the way it seems as we observe skinny people clomping around town in oversize sneakers, getting skinnier, carrying weights, and looking pained in the process. They have that long, lean, painful, constipated look as they lope around the streets in agony. They do not look happy.

Some walk, some jog, others run. Don't ever get in the way of these thin people; they are thoroughly dedicated and concentrated. They'll run right over you as if they wore blinders and were bent on winning the Olympics. They even exercise in the heat at noon rather than eat.

It's difficult to talk to them because they're always timing or counting or taking their pulse to be sure it has revved up to 220 minus their age times 65%. It is very dangerous to interrupt.

We tried to interview one Twiggy-looking person while running along side her. She was huffing and puffing and counting.

"Excuse us, but could you tell us why you are torturing yourself like this?" we asked. No answer, but we did get a pained, don't-bother-me look. She went right on counting. She obviously did not wish to be disturbed. We left her at 147,654 and counting, but we don't know what. How long can these people survive like this?

Sophia Loren has been described as the thinking man's Twiggy. We wondered where all the fat people were. It turned out they were all inside eating and enjoying themselves in cool, air-conditioned comfort.

We interviewed one rotund gentleman munching on fried chicken, French fries, fried okra, and apple pie à la mode.

"What do you do for exercise?" we inquired.

"Eat," he gasped between bites.

"Shouldn't you be out exercising instead of doing this?" we suggested.

He was not counting calories, pulse, or anything else and did not seem to mind at all talking between mouthfuls.

"We all have to die someday, and I want to die happy, right here like this, not out there on the street looking like that," he said, laboring to get some more breath. We felt greatly relieved to get another viewpoint.

Dieting is another popular craze these days. Slimfast, birdseed, low-cal this and that, low cholesterol, raw vegetables, raw fruit, salt free, fat free, all very unappetizing and unappealing. People live on this stuff. Whatever happened to fried mullet and plenty of mashed potatoes slathered in thick, creamy gravy? Gone!

Fat can't be all bad. Look at Oprah Winfrey and Rush Limbaugh. They're not suffering. They're the epitome of success, and they're still fat although Oprah goes up and down like a Yo-Yo. Mostly up. The rate of recidivism for dieters is high. Can you imagine how mean Rush would be on a diet? He's mean enough already. He eats live liberals. (He is now on a diet.)

President Clinton has taken a moderate, middle-of-the-road approach: he runs to McDonald's and eats a hamburger and French fries. We don't think this will work.

Well, to each his or her own. Some people dwell on health. They suffer through endless hours of torture and unhappiness. "No pain, no gain," they say. Others just sit back and relax and eat and get fat and die young.

You have two alternatives: run yourself to death or eat yourself to death, whichever comes first.

Buckle Up

> Golf is a good walk spoiled.
> *Mark Twain*

Golf used to be a sport where you walked around the course following the ball for exercise. Clubs had wooden shafts and wooden heads. Woods were wooden. Metal heads were teensy scraps of iron, so small you might miss the ball completely, like waving a wand. Not any more.

Irons are now the size of horse shoes with all kinds of improvements to keep you from hooking or slicing. (They don't really work.) Woods are now metal, exotic metal to make the ball go farther, faster, and straighter. (They don't really work.)

Golf balls now have more dots and all sorts of new insides to make them go high or low, as desired, and straight. (This is just so much baloney.)

Golfers no longer walk. They ride. Arnold Palmer calls this polo. Players chase the ball in Cadillac-style golf carts equipped with TV, computers, and distance-finding devices. No more guesswork on how far to the hole. Unfortunately nobody has thought to make the hole bigger. This would be a big improvement. As Winston Churchill remarked, "Golf is a game of putting a small ball in an even smaller hole ...."

The government has now stuck its nose into the game. Some Congressperson has decided to require seat belts on golf carts, believe it or not. This law would go to OSHA, the government's office on health and safety in the workplace, for enforcement. These are the same bureaucrats who require employees to go to the bathroom. Victims refer to them as Our Special Hell Administration.

When golfers find out about this new proposal, they are going to be fighting mad. Imagine buckling and unbuckling a seat belt every time you catch up with your golf ball. That would be 100 times for us hackers. You could wear out your belt and your lap in one round. We can imagine OSHA inspectors lurking behind every bush ready to pounce on violators.

We questioned several golfers standing on the first tee. "What about seat belts?" They turned beet-red, then they threw down their hats and stomped them into the ground. Then they threw their clubs into a tree. We left—it didn't seem safe around people in uncontrolled rage.

We did contact club professional Woody Woods who was trying on a seat belt. "Is this necessary?" we asked Woody.

"Are you kidding? How many head-on golf cart crashes have you seen? Have you ever seen a cart tip over? Seat belts are dangerous. If you drove into the lake, you might go down with the ship. OSHA is nuts. Congress is nuts."

"But shouldn't we be on the safe side just in case of a wreck?"

"Hell no, next thing you know they'll make golfers wear hard hats and armor. You can't play golf like that. What will they think of next? Seat belts on lawn mowers or airbags on wheelchairs?"

He took the seat belt and wrapped it around our neck. We got the message.

Sproing!

> I love sports. Whenever I can, I always watch Detroit on the radio.
> *President Gerald Ford*

Basketball is a game for giants. Now that the hoop season is over, we have pondered what to do about these Paul Bunyans of the hardwood.

The problem is this: the basket is too low. In the good old days the players shot from underneath the basket. Now these long-legged behemoths shoot from up above looking down. Even we could score that way.

The idea to raise the basket actually came to us from our 10 year old granddaughter. She plays basketball in a youth league, and their rules *lower* the basket for kids that age, who are too short to shoot at a backboard at regulation height.

Fair is fair. The Equal Opportunity Act demands it. It you lower the basket for the short ones, you certainly ought to raise it for the tall ones who loom above the basket and plunk the ball down through the hoop. Like dropping apples into a barrel. You can't miss.

We decided to try the idea out on the Sopchoppy Pulpwooders, a professional team led by veteran coach Pott Shott.

"How tall are your players?" we asked Pott.

"Oh, Lord, way up there. Forwards 8 feet or more, a center over 10 feet. On a cloudy day I can't see their heads."

"What do you think about our idea of raising the basket for these freaks?" we asked.

"Well, that depends. We have some short guards who wouldn't be able to reach the basket. They'll be mad as hell. What do I do about them?"

We suggested pogo sticks for those short people. Pott agreed to try this out and arranged a practice. It was a disaster.

The guards bounced around like kangaroos. They kept "traveling" and "double-dribbling" because the frequency of the dribbles didn't match the frequency of their pogo sticks. One short player was right under the basket when his pogo launched him on its up cycle. His head went up through the rim of the basket, and he got stuck in the net. We cut him down, but he was not pleased.

"This ain't gonna work," said Pott.

"How about letting these short guys carry stepladders?" we wondered.

"I'm game," said Pott, "let's try it."

This worked pretty well for a while with the guards carrying light aluminum stepladders. But then when one short guard got close to the basket, put down his ladder, and climbed up to shoot, one of the monsters crashed into the ladder and catapulted the guard out into the audience. He tried to shoot from the reserved seats but missed.

"This will never do," said Pott.

"What'll we try next?" we asked.

"I'll tell you what. We'll make the tall ones shoot from at least 100 feet. That way they won't get an unfair advantage. They won't get in the way of the guards."

This system worked perfectly. The tall people looped long, high, 100-foot shots and missed, and the guards scored all the points for a change. The short people love this idea.

Basketball is serious business. We are forwarding the results of our research to the National Basketball Association. We think they will be interested.

From Left Field

> Fans, don't fail to miss tomorrow's game.
> *Dizzy Dean*

Spring has sprung along with baseball. Spring training is here even before basketball season ends.

Hockey runs until summer when the ice melts. Arena football takes us to summer football training camps and pre-season gridiron games. The baseball world series carries over into football season. Then the cycle starts all over again; there is no end to it.

Charlie Somerby is in his glory in front of his T.V. Baseball is back, and he seems to have the sports page pretty well covered; but we would like to sneak in a little baseball history just to get a word in edgewise.

Our two favorite baseball heroes are Casey Stengel and Yogi Berra. These guys played and then managed. They are also comedians.

There is the story about Casey who was coaching a rookie kid in how to play right field. Casey hit the rookie some balls. They went over the kid, around him, through him, and every which way. Finally Casey gave up and said, "Here, kid, let me show you how to play right field."

Stengel put on a glove and went to right field and missed every ball hit to him. Finally he gave up in disgust, threw down his glove, stomped on his hat, and stormed back to the dugout. "Son," he said, "you've messed up right field so bad nobody can play it."

Casey was also famous for his speeches; he was the master of the garbled word. Following is a portion of a speech he gave on how he managed his team:

"No manager is ever going to run a tail-end club and be popular because there is no strikeout king that he's going to go up and shake hands with and they're going to love you because who's going to kiss a player when he strikes out and I got a short-stop which I don't think could have been a success without him.

"If you mix up the in-field, you can't have teamwork. It's a strange thing if you look it up that the Milwaukee Club in the morning paper lost a double header and they got three of my players on their team and you can think it over." (Taken from *The 776 Stupidest Things Ever Said.*)

What did Casey say? Maybe Charlie can translate it. But Casey was a winner, with a great gift of gab.

Of his first baseman Gil Hodges, he once remarked, "He is so strong he could snap your earbrows off."

Of mortality, he said, "A lot of people my age are dead at the present time."

He once asked Yogi Berra, "What would you do if you found a million dollars?"

Said Yogi, "I'd find the fellow who lost it, and if he was poor, I'd return it."

Next time we'll listen to Yogi Berra's words of wisdom.

Ear Today, Gone Tomorrow

Just because you're paranoid doesn't mean they aren't really after you.
Anonymous

Boxing started with the cavemen, and the cavemen are still boxing in Las Vegas, Nevada.

We just witnessed a so-called $60,000,000 prize fight between two human behemoths, Mike Tyson and Evander Holyfield, for the heavyweight championship of the world. Prior to their primetime main event two women boxed and bloodied each other. What is the world coming to?

Tyson and Holyfield will split the pot and get $30,000,000 each. At least Tyson was supposed to until he bit off part of Holyfield's ear. Unbelievable. The Marquis of Queensberry rules don't cover biting. Biting is a two-year-old stunt, not for 31 year olds. Tyson may have learned that little trick during his three-year sabbatical in the penitentiary.

We felt compelled to investigate this amazing feat so first we talked to Mr. Tyson.

"Isn't biting off an ear ungentlemanly, even in boxing?" we asked.

"You don't understand. Evander and I are great friends. That was just a tender little nibble to show him how much I care. I didn't get the whole ear, just a hunk," explained Tyson. "That's true love."

"Well, what about the other ear? You bit that one too," we reminded him.

"That's just another way of showing how much I love Evander."

"What about Holyfield? Did he bite you back?"

"Naw, but he butted me with his head. That's against the rules. It's mean."

"How did the ear taste?"

"Terrible. It might have been OK with a little ketchup, but I had to spit it out. I was afraid I couldn't digest it."

"Did you give him his ear back?"

"I dunno. Last I seen of it was laying on the floor."

"Why were you disqualified?"

"It was that dumb ref. He don't understand nothing about love."

We were not at all satisfied with these answers so we questioned Evander Holyfield, winner and still world's heavyweight champion. He was bleeding.

"How's your ear?" we asked.

"What's left of it is OK. We can't find the rest of it." There were people crawling around on their hands and knees in the ring looking for the ear.

"Maybe Tyson swallowed it," we said.

"If he did, I hope he chokes. Maybe he'll get indigestion or distemper or something."

"What will you do if you find it?"

"I want to frame it and send it to the Smithsonian Museum, but the doc claims he can fasten it back on. My wife don't want me with half an ear. She's mad as hell."

"Are you mad at Tyson?"

"Yeah. Look at me. I look like I been through a shredder."

"He says you butted him with your head."

"Baloney. That would be like butting a concrete wall. See, here, look, my head's OK. No holes in it."

"Mike says he loves you."

"He does? Well, that's different. I was afraid he was mad at me."

"Are you guys going to kiss and make up?"

"I sure am, right after I sue him for $30,000,000. Now help us find my ear."

The Nevada Athletic Commission threatens to withhold 10% of Tyson's purse. That would only leave him $27,000,000. That's a pretty good night's work for two ears in just eight minutes.

That says a mouthful about boxing.

Mayhem

> Football is one of the highest forms of spiritual exercise.
> *Rt. Rev. W.T. Manning*
> *Circa 1926*

Football madness is upon us again. Football is organized mayhem. As Minnesota Viking player Joey Browner observed, "It's not good for business if you care for a second whether blood is bubbling from a guy's mouth."

Years ago actor Andy Griffith, who started out playing a North Carolina cracker, came up with a hilarious phonograph record about seeing his first football game. The name of the record: "What It Was Was Football." To paraphrase Andy: "It looked like some kind of a contest on a little cow pasture with stripes on it where a bunch of them mean boys run up and down the field from one end to the other with a funny lookin' kinda bundle, all the time trying not to drop it and tryin' to keep from steppin' in something. As fast as they hurt one of them boys and run him off, they run another one on." Little did he know!

So here we go again into the gladiatoriums for the fall football spectaculars. Football is serious business to many, but fortunately there are a lot of football laughs as reported in our favorite text book, *The 776 Stupidest Things Ever Said*, collected by Ross and Kathryn Petras (Doubleday, 1993).

For example: "If there's a pileup, they'll have to give some of the players artificial insemination." This was Curt Gowdy, TV announcer, during an AFL all-star game on a field flooded by heavy rain.

One of our favorite coaches of all time was Florida State's Bill Peterson: "They gave me a standing observation."

"Please lead us in a few words of silent prayer."

"I'm the football coach around here and don't you remember it."

"This is the greatest country in America."

"We're not going to be any three-clouds-and-a-yard-of-dust team."

Continuing on with other famous football statements:

"I want to gain 1500 or 2000 yards, whichever comes first." George Rogers, New Orleans running back.

"The odds that we won't win a game this season are 999 out of a hundred." Anonymous football coach.

"Yale football keeps the alumni sullen but not mutinous." Coach Herman Hickman.

Many great tales abound about football coaches. There is the famous tale about Duffy Daugherty, Michigan coach, who had a terrible losing season. He got sick and had to go to the hospital. The board of regents sent him a sympathy telegram: "We hope you get well by a vote of 9 to 8."

Bring on the gladiators!

Divot Diggers

> Nothing increases your golf score like witnesses.
> *Anonymous*

This area is loaded with golf courses and rabid golfers.

Here comes the annual senior pro golf tournament at the Moors. This is a great attraction for golf nuts and others in the neighborhood.

One of the beautiful things about golf is the number of great golf stories. We collected a few of them. Golf nuts will appreciate the eternal truths in these stories.

There is the one about the two elderly golfers who arrived in Heaven to find the most beautiful golf course they had ever seen. Everything was perfect.

Said one, "Man, where have we been all this time? Look what we've been missing!"

Said the second, "Yeah, just think of it. We could have been here two years earlier if we hadn't eaten all that damned oat bran."

Lightning once struck pro golfer Lee Trevino on the golf course. That scared a lot of golfers and spectators alike. During a later lightning storm, another golfer held his 2-iron up over his head. His partner was horrified.

"For God's sake, don't do that. Lightning might strike you."

"Don't worry," said the golfer, "not even God can hit a 2-iron."

Even later God himself was playing golf. When he hit his ball into a lake, he walked across the water to find it. This act really impressed a spectator. "Look at that guy; he thinks he's God."

"No he doesn't," remarked a friend, "he thinks he's Jack Nicklaus."

Then there are some cruel golf stories. One golfer was standing on the 10th tee when a hearse and funeral procession passed by. He took off his hat, held it over his heart, and stood at attention.

"What's that all about?" asked his partner.

Said the golfer, "That's my wife's funeral, and the least I can do is pay my respects."

Another amateur golfer took lessons from the club pro. During one lesson he sliced a golf ball far to the right. The ball hit a house, broke a window, and killed a baby.

"Oh my God," moaned the poor distraught golfer, "what should I do?"

Said the pro, "Try keeping your left arm straighter."

A nicer story is the one about the golfer who was shipwrecked and then washed up on a deserted island. Lo and behold, a beautiful girl appeared. She served him cocktails and a scrumptious dinner. Afterward she asked, "Do you want to play around?"

The golfer was flabbergasted. "Good Lord, do you mean to tell me there's a golf course here, too?"

So much for golf stories. But even the great Winston Churchill could not resist commenting on this popular sport: "Golf is like chasing a quinine pill around a pasture. It is a game of putting a small ball into a smaller hole with instruments ill-designed for the purpose."

So be it. There are millions of us crazy enough to try. But Phyllis Diller has the last word: "Golf has taught me there is both pain and pleasure in the game. Golf spelled backward is 'FLOG.'"

Fore!

To Each His Own

> I'd like you to know I have the body of a man half my age.
> Unfortunately he's in terrible shape.
> *Heavyweight Champ George Foreman*

There are all kinds of sports for all kinds of athletes. Some athletes pump iron and run and diet; others take a more sedentary approach to athletics. We thought we would review some of the possibilities.

Jogging is big for real jocks. One of our friends went to the doctor to see how he could improve his health, attitude, and family relations at home.

"You need to jog 10 miles a day," advised the doctor.

Two weeks later the man called the doctor to report in.

"How's it going?" asked his doctor.

"I don't know," said the jogger, "I'm 140 miles from home."

Tennis is another active sport. Supreme Court Justice Hugo L. Black was an avid tennis player. Said he, "When I was 40 my doctor advised me that a man in his forties shouldn't play tennis. I heeded his advice carefully and could hardly wait until I reached 50 to start again." So you're never too old to exercise.

Golf is less active. The way Mark Twain looked at it, "Golf is a good walk spoiled." Most golfers now ride carts.

One golfer's ambition was to live to be 100 so he could shoot his age. Famed pitcher Satchel Paige, who still pitched at a ripe old age, didn't let the years bother him. "Age is mind over matter," he said. "If you don't mind, it doesn't matter."

Even ping-pong can provide a workout. The Chinese are masters of this sport. With over a billion Chinamen, women and children, one jokester laughed at the idea that ping-pong was their only indoor sport.

Then there are those other sports where you don't have to strain a single muscle. As one example, consider horse racing. As one wag advised, "One way to stop a runaway horse is to bet on him."

One big bettor was always looking for hot tips or signs to help him bet on the horses. One morning he woke up at 3:00 a.m. The temperature was 33°, and it was the 3rd of December. Then he discovered that a horse named Trio was running in the 3rd race. "Boy, this is my lucky day," he enthused. He bet everything he had on Trio to win. Sure enough, Trio ran third.

Fishing is another sport which doesn't provide much exercise. Mostly you sit, and very still. Two fishermen were sitting in the boat. One was lighting sticks of dynamite and tossing them into the water and was producing lots of fish. Finally the second man spoke up. "I'm a game warden, and that's illegal."

The first man lit another stick of dynamite and held it in his hand. "All right, so you're a game warden. Are we going to fish, or are we going to talk?"

In colder climes ice fishing has become a popular sport. Not much exercise after you chop a hole in the ice. On one occasion a young boy was catching all the fish among a large crowd of fisherpersons. The others couldn't stand it and finally asked the boy what his secret was.

"Uh, uh, uh," the boy mumbled.

"Speak up, boy. We can't understand you. We want to know how you do it."

The boy spit a large wad out of his mouth. "I just keep the angleworms warm."

Hunting is a sport which can be active or inactive. For those who don't relish tromping through the woods or swamps for game, there is fox hunting where the crowd just sits in a stand and watches. In one such case a pack of hounds pursued the fox; the dogs were yelping and running hard. By mistake, a female hound in heat got mixed in with the pack. As the hounds crossed an open field, one observer called out, "How's the fox doing?"

Came the answer, "He's running fifth!"

So much for athletics. Take your choice, active or inactive, but be a sport.

Looking Down

> When small men cast a long shadow, the sun is going down.
> *Venita Cravens*

The country is fast running out of role models.

We used to use the president, but that doesn't work any more. We have the current White House resident saddled with Misses Lewinsky, Willey, Jones, and Flowers, among others, and with the persistent Judge Starr. Nineteen and counting of Mr. Clinton's Arkansas and administration henchmen have already either been indicted, convicted, or jailed. More to come. So much for that crowd of role models. Then there's Al Gore and the wealthy nuns.

So we have to turn away from political role models and look to athletes. Kids look up to jocks. The picture is not much better there.

Remember Mike Tyson and his rape case? Then he bit off Evander Holyfield's ear. Mike is out of jail, banned from boxing, and trying out as a wrestler, a grand role model sport. He does still have his millions to look up to. (He has just been reinstated.)

Next comes a basketball character named Sprewell who attacked his coach and threatened another player with a 2 x 4, among other crimes. He's out of basketball temporarily, hopefully permanently.

Then there was a fine, upstanding, heavy-hitting, highly paid baseball player who spit in the face of an umpire who displeased his majesty. The umpire threw him clear out of the park, but he's right back inside slugging baseballs after a short sabbatical and reprimand. We pay dearly to go see these miscreants play. Do we have to look up to them, too, even if they make millions for spitting?

We have talented athletes striking or holding out for more millions; there is big league free agency and lots of other good union benefits. Pity the poor school teachers, police, and other useful, productive citizens who have to hold out for a whole lot less.

Comes now the just finished saga of the U.S. hockey team at the winter Olympics in Japan. Instead of sending amateurs to play in the true spirit of the games, we shut down the professional National Hockey League and sent the pros to Nagano to represent the U.S., Canada, and other nations.

Unlike our glorious winning U.S. Olympic amateur hockey players of the glorious past, these pros failed miserably. The U.S. team didn't even make the playoffs. Afterward the shamefaced U.S. pros threw a party to celebrate losing. They got drunk in their living quarters and apparently shot out the lights with pucks. They threw chairs, players, etc., out the windows and generally trashed the joint to make up for not winning. We guess they were practicing getting sent to the penalty box where they should go permanently.

They showed the world that, after all, they were real, tough, glorious pro losers. Kids are supposed to look up to these sorry characters?

Compare their behavior to that of hockey immortal Wayne Gretzky whose team didn't win a medal, either, but who took it like a man, graciously, without throwing a tantrum. Or Michelle Kwan who had to settle for second place in figure skating.

Where else do we go for role models? Thank God for Gretzky, Kwan, and our own Emmitt Smith and Roy Jones, Jr., and maybe, some day, another president.

Red-Hot and Thumb Screws

> Only two kids enjoy high school. One is the captain of the football team.
> The other is his girl friend.
> *Letter to Ann Landers*

More education scandals abound.

Parents are accusing some high school football coaches of beating up on their student athletes by punching, poking, cursing, slapping, head butting, and so on. This is called discipline.

Other sports fans, quarterback clubs, jocks, ex-jocks, and even some parents support these coaches. They stick up for pugnacious discipline. They claim that coaches are different from other teachers as they have to be extra-tough in the sports world.

We have to put this matter in proper perspective. After all, athletics are first priority, far more important than academics. The idea is to win, not necessarily to learn or to graduate.

We strongly endorse strict physical discipline for football players and have some recommendations of our own. Mere pushing, punching, and cursing are not always severe enough.

For instance, what do you do with a fullback who keeps fumbling the ball? No punishment is too severe. We recommend hanging a fumbling fullback upside down while holding a football to see if he can hang on. If he drops the ball, leave him there permanently. This cure is miraculous. Just find another fullback, one who eats Elmer's Glue.

How about the runningback who is too slow? How do you speed him up? We recommend a liberal dose of RED-HOT liniment in his jock strap. This incentive works wonders. He can virtually fly around left end without touching the ground.

Quarterbacks are a problem. There once was a quarterback who couldn't remember the plays. He needed something to clear his mind. We suggested a blast of compressed air into his helmet while he's wearing it. This treatment could blow his brains out, but what good are brains if he can't remember the plays? So get another quarterback.

What about a wide receiver who can't catch the ball? The answer is twofold: tighten his thumb screws during practice and then hang him up by his thumbs overnight. This remedy should extend his reach.

These are relatively mild disciplinary measures. In more severe cases, such as missing a tackle or running the wrong way, we suggest caning, far more effective than punching a player in the nose.

What about a losing team? If coaches don't win, the fans get hysterical. A losing team should suffer the ultimate indignity: all the players should be locked in stocks on public display in front of the quarterback club for a month. This treatment is known as attitude adjustment and brings shame, disgrace, and humility to wimps who lose.

What about disciplining coaches? We suggest the RED-HOT treatment. They will disappear into the sunset at a high rate of speed.

Berra-isms

Good hitting always stops good pitching, and vice versa.
Casey Stengel

Yogi Berra is baseball's favorite stand-up comic. Yogi is the master of the one-liner.
Nobody is exactly sure how many of these quips were actually his, but many of the best are attributed to him anyway because of his reputation for comedy.
The following Yogi Berra-isms appeared in the authoritative research text *The 776 Stupidest Things Ever Said*:

Yogi - "You mean you get seasick?"
Rube Walker - "Do I ever!"
Yogi - "On water?"

Or this one:

Tom Seaver - "What time is it?"
Yogi - "You mean right now?"

Then there are dozens of Yogi one-liners on baseball. Take your pick:

"Mantel's a switch hitter because he's amphibious."
When asked his cap size - "I don't know. I'm not in shape yet."
"He's a big clog in their machine." (talking about Ted Williams)
"Slump, I ain't in no slump, I just ain't hitting."
"I got a touch of pantomime poisoning." (explaining to Casey Stengel why he couldn't play)
"How can you think and hit at the same time?"
"He can run any time he wants. I'm giving him the red light."
After a bad season - "I wish I had an answer to that because I'm getting tired of answering that question."

But his comments didn't just relate to baseball. For instance:

"A nickel ain't worth a dime any more."
"If I didn't wake up, I'd still be sleeping."
"Steve McQueen must have made that movie before he died."
After he had seen a streaker and was asked if the person was male or female - "I don't know, they were wearing a paper bag over their head."
"I usually take a two-hour nap from one o'clock to four."
"I'm wearing these gloves for my hands."
"It was hard to have a conversation with anyone, there were so many people talking."
Elderly woman on a hot St. Petersburg day in Florida - "Good afternoon, Mr. Berra. My, you look mighty cool today."
Yogi - "Thank you, ma'am. You don't look so hot yourself."

Then there was the time Yogi was on the phone to Joe Garagiola who was lost.

Yogi - "Where are you?"
Joe - "Some guy says to tell you I'm at the library."

Yogi - "Oh, you ain't too far, just a couple of blocks. Only don't go that way, come this way."

Yogi finally summed it all up - "I really didn't say everything I said."

Said his son Dale, "The similarities between me and my father are different."

Apparently not!

Controlled Riots

> If you can't make the putts and can't get the man in from second on the bottom
> of the ninth, you're not going to win enough football games in this league,
> and that's the problem we had today.
> *Sam Rutigliano, Coach of the Cleveland Browns*

Well, if you don't know whether you're playing golf or baseball or football, no wonder you lose. But as Dan O'Day says, "Pro football is a tough game. The linebackers take hostages."

Win or lose, football *is* a rough sport at all levels, and the season's head-knocking and bone-splintering are already crashing full speed back and forth with ups and downs on thousands of gladiatoriums throughout the land. As well as mayhem, funny things happen on the way to the goal posts. We like to collect football stories whenever we come across an interesting one.

The Army-Navy game is a perennial rivalry over scores of years. Army does not love Navy and vice-versa. On one occasion Army called a time-out with just minutes left to play. Navy mistook this lull for the end of the game and retired to their locker room. Army stayed on the field and resumed play without Navy on the field. Navy swore it took Army four plays to score.

There is the story about a game where one team took a severe battering on the very first few plays. The opposition's defense was literally cutting the offensive players in half with vicious tackles.

So far the offense's star fullback, Calhoun, had not carried the ball. The demoralized fans in the stands began to chant, "Give Calhoun the ball. Give Calhoun the ball."

Finally the frustrated quarterback stood up in the huddle and shouted to the crowd, "Calhoun say he don't want the ball."

The poor coaches continually catch hell if they don't win. Fans are unforgiving. Even the players complain. Firings are numerous.

We heard of one case where the coach was sweating out a game which was still tied during the last desperate minutes. Then he lost both his first-string and second-string quarterbacks to injuries. All he had left was a distant third-stringer.

Now hoping at least to preserve the tie by stalling until he could run out the clock and end the game, he summoned his third-stringer from the bench for instructions. "Son," he said, "it's all up to you. Don't do anything stupid. Don't try to be a hero; I just want to save this tie. You just go in there and hold on to the ball as long as you can. We're on our own ten yard line. Do three quick quarterback sneaks and down the ball as fast as you can. Then on fourth down punt the ball as far as you can."

So the quarterback went in and ran the three sneak plays. He gained 85 yards and then punted on the 5-yard line.

Naturally the coach was mortified. After the game he questioned the quarterback, "Son, you did a great job of running. We could have won the game. Why on earth did you kick on their 5-yard line?"

Said the young gridster, "Because we have a dumb coach."

So there you have it. Pity the poor coaches. They can't win for tying.

Bloop

He fakes a bluff.
Ron Fairly, NY Giant baseball announcer

Fall has fallen upon us, and the World Series can't be far behind. We can just see ace sportscaster Charlie Somerby warming up pitchers in front of his TV.

In baseball a blooper is a pop-up fly ball that lands just over the infield where nobody can catch it. There is also another kind of blooper in baseball: a foot-in-mouth blunder made by a radio or TV announcer.

In studying our favorite reference work, *The 776 Stupidest Things Ever Said* by Ross and Kathryn Petras, we discovered more epic bloopers by some epic sportscasters. Here goes:

One of the most famous announcers was Jerry Coleman who broadcast for the San Diego Padres. He was not immune from blooperitis.

On one occasion, he led off the game with "Hi, folks, I'm Jerry Gross. No, I'm not, I'm Jerry Coleman." That was bad enough. It gets worse:

"And Kansas City is at Chicago tonight, or is that Chicago at Kansas City? Well, no matter. Kansas City leads in the eighth, 4 to 4."

"There's someone warming up in the bull pen, but he's obscured by his number."

"Now Juantorena opens his legs and really shows his class."

"Young Frank Pastore may have pitched the biggest victory of 1979. Maybe the biggest victory of the year."

"Pete Rose has 3000 hits and 3014 overall."

"He slides into second with a stand-up double."

"McCovey swings and misses, and it's fouled back."

"If Rose's streak was still intact, with that single to the left the fans would be throwing babies out of the upper deck."

"Rick Rolkers is throwing up in the bullpen."

"There's a hard shot to LeMaster and he throws Madlock into the dugout."

"They throw Winfield out at second, and he's safe."

"Winfield goes back to the wall. He hits his head on the wall and it rolls off. It's rolling all the way back to second base. This is a terrible thing for the Padres."

"Eric Show will be 0 for 10 if that pop fly comes down."

As if all those weren't enough, comes now renowned announcer Curt Gowdy who fields his own share of bloopers:

"Folks, this is perfect weather for today's game. Not a breath of air."

"(Pitcher) Luis Tiant throws from everywhere except between his legs."

"Rex Morgan winds up his career today, the only starter in the starting lineup."

"The Baltimore Colts are a bright young team. It seems as if they have their future ahead of them."

Of course, the most famous baseball blooperite of all time was the beloved and incomparable Yogi Berra. Those guys are full of bloops, and we're the bloopees.

Foul Balls

> Baseball may be the nation's game, with throw and run and hit;
> but first among the skills it seems, is learning how to spit.
> *Art Buck*

Here is March already. Spring is springing so baseball, spring training, and Charlie Somerby can't be far behind.

Our favorite reference book, *The 776 Stupidest Things Ever Said* by Ross and Kathryn Petros (Doubleday 1993), includes some baseball wisdom in honor of the occasion.

The greatest one liner in baseball, in fact in the whole world, is the renowned and revered Yogi Berra. Here is just a sampling of his greatest works:

"It never happened in a World Series game, and it still hasn't."

"I'll get it when I die," explaining why he bought life insurance.

"Nobody goes there anymore. It's too crowded."

"If you ask me anything I don't know, I'm not going to answer."

"It was hard to have a conversation, there were so many people talking," while dining at the White House.

The great pitcher Dizzy Dean, later a play-by-play announcer, was famous for his boo-boos during St. Louis games:

"The commotion in the stands has something to do with a fat lady. (Pause). I've just been informed that the fat lady is the Queen of Holland."

Curt Gowdy, notable sports announcer, had a knack for throwing out mixed messages:

"Brooks Robinson is not a fast man, but his arms and legs move very quickly."

Clark Griffith evaluated his pitching staff: "The fans like home runs, and I have assembled a team of pitchers to please them."

When the fans hung their coats on the outfield fence, the announcer got on the air: "Would the fans along the outfield please remove their clothes."

Said player Jim Gantner when he forgot to appear on a talk show: "I must have had ambrosia."

Proclaimed manager Don Zimmer on his team's 4-4 record: "It just as easily could have gone the other way."

Admitted manager Sparky Anderson: "Our pitching could be better than I think it will be."

Announced New York Mets broadcaster Ralph Kiner: "Today is Father's Day so everyone out there - Happy Birthday!" Later Kiner said, "Some quiet guys are inwardly outgoing."

"It could permanently hurt a batter for a long time," explained Pete Rose about a batter hit by a pitched ball.

Yankee Doug Mattingly described pitcher Dwight Gooden: "His reputation preceded him before he got here."

A Texas Ranger apologized: "People think we make $3-4 million a year. Most of us only make $500,000."

"Last night I neglected to mention something that bears repeating," said a San Francisco Giant announcer.

Play ball!

ON TRAVEL

If you come to a fork in the road, take it.
Yogi Berra

No unmet needs exist...current unmet needs that are being
met will continue to be met.
Transportation Commission, Mariposa County, CA

When two trains approach each other at a crossing, they shall both come
to a full stop, and neither shall start up until the other has gone.
Kansas Law

I debated with myself the night before I left for Russia whether 2 extra pairs
of socks would constitute capitalistic affluence.
Will Rogers

Boom

HEADLINE: Commode Explodes in Wadley, Alabama.
by Associated Press

This dramatic news signaled a week of tragedy in poor Wadley, Alabama. Here's what happened:

The trouble started in City Hall where the mayor and a policeman got into an argument over an alcohol-related arrest.

"He called me a liar," protested the mayor. "He pushed me backwards."

The policeman filed a complaint, and the sheriff sent two deputies to arrest the mayor. The city police chief drove the mayor to jail.

There was an altercation. Nobody can agree on what actually happened there, but the sheriff was charged with assault and was arrested by the coroner.

Then the commode in City Hall blew up. Sitting on the commode at the time was the police chief who disappeared along with the commode, bathtub, plumbing, electrical fixtures, and the west wall. He has not been located as yet. We do know he was airborne.

The chief's widow naturally suspected the sheriff of perpetrating this foul deed. She pressed charges, sued the sheriff, filed for the chief's life insurance, asked for probate of his will, and raised all kinds of hell.

Meanwhile, while passing a Delta airliner, the chief did manage to wave at the passengers. We have had no further word. NASA has been alerted, and the astronauts have been told to be on the lookout to avoid a collision. One UFO in Gulf Breeze, Florida, was mistakenly identified as the chief. There have been no further confirmed sightings since Delta Flight 478. We wish we had better news.

Excess Baggage

> Airline Ticket Agent: Where do you want to go?
> Traveler: I want to go to the same place my bags are going.
> *Anonymous*

We recently visited the new Denver airport to see how the new baggage handling system there was working out. You may have read that there were problems. In fact, these problems delayed the opening of the new airport, costing Denver $1,000,000 per day. During the delay, the new runways started cracking up. It may just be cheaper to abandon the whole airport. The loss would be relatively minor, only a billion or two.

Since finding your own luggage is always a scramble after any flight, we thought perhaps we might salvage the system for use at our fine new airport. We always like to stay on the cutting edge of technology.

In order to get the lowdown on the Denver baggage system, we consulted with Mr. Timothy Tossit, chief engineer with the architectural firm of Throwit, Stompit, Pitchit, Breakit, and Leavit. He was most cooperative.

"This is the very latest word in the baggage industry," he bragged, obviously pleased with himself.

"What are some of the advantages?" we wanted to know.

"Well, first of all, we incorporate a baggage testing feature. You've seen Samsonite drop luggage off of tall buildings to prove how tough it is. Our system is much simpler and cleaner. A robot hurls the bags against a concrete wall at 150 m.p.h."

Tossit was ecstatic about his invention. The air was filled with flying luggage. The operator crouched inside a concrete shelter. As we watched, the robot whipped one bag through the front window of the terminal and into the parking lot toward an unsuspecting skycap.

"Oops!" said Tossit. "Another new feature is curb service. No standing in line. The last and most important advantage is our suitcase dismantling feature. The robot empties your bag right here in the lobby, handy for customs inspections or looking for your toothbrush."

The terminal floor was littered with drawers, bras, deodorant, and shaving cream. Passengers sorted through the wreckage with rakes trying to identify their belongings. A man from the salvage company shoveled up the skeletal remains of suitcases.

We have made a report on this remarkable system to the local authorities.

In the meantime, avoid Denver.

If somehow you can't and end up standing in the middle of the new Denver airline terminal, *duck!*

The Pecking Order

> A bird in the hand is safer than one overhead.
> *Anonymous*

Various malfunctions have grounded NASA's space shuttles from time to time: weak "O" rings, leaks, computer failures, weather, etc. The latest malfunction was two woodpeckers.

While practicing on the ground, the astronauts heard a strange pecking sound. The head astronaut looked out the window and discovered a pair of woodpeckers trying to peck holes in the shuttle. He reported immediately to mission control. "This won't fly," declared the mission director.

The government called in a panel of experts to examine the problem. The group included marriage counselors, welfare workers, social workers, ornithologists, bird watchers, bird callers, birdseed salesmen, foresters, telephone linemen, the National Rifle Association (NRA), Greenpeace, environmentalists, ornithological psychologists, and Vice-President Al Gore. While the experts deliberated, the woodpeckers ate more of the shuttle. The holes grew more expensive daily, millions in damages.

The NRA wanted to shoot the birds and get it over with in a big hurry. The Woodpeckers' Rights Association protested vehemently. Other panel members proposed every conceivable tactic to lure the birds away from the shuttle before it disappeared completely.

The power company erected a telephone pole as an alternative. The birds ignored it. The bird callers tried mating calls. The birds wouldn't even listen; apparently they were already engaged. Birdseed companies concocted gourmet meals. The birds spit them out and kept right on pecking. Social workers and psychologists offered counseling. The Vice-President tried talking to them. Nothing worked.

Finally NASA removed the shuttle from the launch pad to repair the damage. The woodpeckers were furious. Environmentalists were concerned that the pair might suffer irreparable mental harm if they were left homeless.

The expert panel finally conceived a very simple solution. They stuck a long wooden pole, with predrilled holes in it, on the nose of the shuttle to provide public housing for the woodpeckers. They fastened a Greenpeace member to the pole to chaperone the nesting pair.

Another historical flight: the first woodpeckers in space, accompanied by Greenpeace.

They're Here, But Where?

> There's a lot of uncertainty that's not clear in my mind.
> *Speaker of Texas House*

UFO (Unidentified Flying Objects) believers recently held their national convention. Our own Gulf Breeze is national UFO headquarters and the primary landing area in the U.S.

Frankly we have always been somewhat skeptical of UFO's so we went to the convention to try and be convinced. Wow! These people are really serious. They believe!

We heard endless tales of sightings, communications, prisoner exchanges, diplomacy, foreign aid, kidnappings, tourist visits, and landings. People there swore they had taken UFO rides in weird round spacecrafts with windows and lights and free cocktails and everything. They went first class!

Some UFO victims were seized bodily and carted off as prisoners. Many of the sightings occurred at UFO parties in Gulf Breeze. The longer and later the parties, the more sightings there were. The landing pads in Gulf Breeze are wearing out with constant takeoffs and landings. The tourist bureau is swamped.

Of an even more serious nature, UFO enthusiasts documented cases of permanent aliens now stationed among us on earth. Some UFO believers swore they are in communication with these unauthorized immigrants or tourists or visitors or spies or whatever they are. This may already pose a serious illegal alien problem.

We came home with the distinct fear that we have already been invaded. If so, the question now is where are these UFO aliens and what are they up to? We are determined to find out.

Some of our neighbors are beginning to look strange and act funny. Could we have UFO aliens living right next door without even knowing it? We set out to canvas the neighborhood.

We have one neighbor who constantly works in his yard doing *his own* yard work. He picks up trash and cuts weeds in the neighborhood. These facts in themselves are highly suspicious although he did look reasonably normal and had two eyes, not just one eye in the middle of his forehead.

We snuck up on him as he was digging around in his bushes.

"What's up?" we asked.

"Just digging around," he said with a silly grin, trying to kick some dirt back into the hole he had dug. This looked pretty sneaky. He looked guilty.

Late that night we snuck back and dug up his hole to see what he had buried. Sure enough, we found packages of $100 bills, his company's financial records, and his accountant. This really was suspicious. He might be one, but then he might be a local businessman. We're not sure which.

Another neighbor lives in a rundown house in a yard grown up in weeds. A scorched area in the back yard really seemed odd. Could this possibly be a launch pad seared by UFO booster rockets? A bunch of us UFO investigators staked out the house to watch for a landing. We had refreshments (for medicinal purposes only) and doctored ourselves frequently.

Sure enough, pretty soon in flew a large, round, house-looking thing. The engines shut down, and a load of little people got out and marched off thru the neighborhood selling Girl Scout cookies. The next morning we couldn't remember whether we had seen aliens or Girl Scouts. We're still hazy on that point, but we continue to suspect that UFO aliens are here. They still could be your next-door neighbor although we haven't definitely located any confirmed aliens as yet.

We worry about how many of these people could already have been elected to public office. That could be the source of some of our current political problems. Keep a sharp lookout, and be sure you know whom you're voting for.

Mars or Bust

A billion here and a billion there...
Senator Everett Dirksen

Our Mars spaceship (the Mars Observer) has disappeared from sight along with a billion dollars worth of time, labor, and materials. That doesn't sound like much money these days, but as Senator Dirksen used to say, "A billion here and a billion there, and pretty soon it adds up to real money." A billion dollars would do more for us here on earth than on Mars. Be that as it may, our space craft is still missing with your billion dollars.

So what happened? Where did it go? Where is it? These are all good questions. You need to know where your money went.

Spaceships are really just big computers. They are practically human. They act like humans. They are programmed to take orders from earth and send back information. In fact they are constantly ordered around like henpecked husbands, "Do this; do that; take this; take that." They are not allowed to talk back to their superiors.

Finally, this routine gets old. The Mars space ship got sick and tired of it all. It was fed up.

Fortunately for it, unfortunately for its parent National Air and Space Administration, the engineers had also built into the computer some capacity to think for itself, just like a human. That did it. That's exactly what happened: the Mars craft discovered this big loophole and took full advantage.

The space craft finally decided, "I've had it. I'm up to here in it. I'm fed up. I'm not going to take it anymore." With that it tuned out NASA in Houston and took off on vacation.

Free from the clutches of earth people, our spaceship decided to go ahead and orbit Mars and take a look around at the scenery. A permanent paid vacation on NASA.

NASA was relieved to get the news; they were also mad. They are still calling Mars and begging the spaceship to tune back in. Fat chance. It's all over. It settled permanently on Mars and married a nice young Russian spaceship.

The Mars settlement is now eligible for U.S. foreign aid. The administration is now busy trying to figure out how to get the money up there to them.

There goes another billion... or two.

Mars or bust. Go for broke.

The UFO's Have Landed...Again

> There are two kinds of truth. There are real truths and there
> are made-up truths.
> *Marion Barry, Mayor of Washington, D.C.*

The UFO's have returned to Gulf Breeze. Just recently there were several reliable sightings.

We trust the sober judgment of Gulf Breeze residents, since it's a dry county. These sightings were made by cold sober citizens.

No sooner do these reliable, trustworthy watchers report the arrival of the UFO's than naysayers in Pensacola start pooh-poohing the whole idea. Since Pensacola is wet, it is obvious the Pensacola killjoys are not to be taken seriously.

The Pensacola wet-blanket brigade put out the silly idea that these UFO's were really blimps which had dropped in on Gulf Breeze for a visit. This idea is preposterous.

Several regular customers at Trader Jon's bar in downtown Pensacola claimed that the craft, whatever it was, had the name "BLOCKBUSTER" written on the side. (These customers obviously had been in the bar all day long.) Certainly this fact has not been confirmed by Gulf Breeze witnesses who were out at night.

Since nobody had ever seen a UFO with a Blockbuster ad draped on the side, we called in Colonel High Flyover, an Air Force expert in charge of UFO's and AWOL's. The colonel stonewalled us.

"The Air Force doesn't believe in UFO's. It was probably just the Navy messing around over there."

He was no help so we questioned a CIA agent who cannot be identified. He spoke to us behind a curtain through a scrambler so that he sounded like a load of gravel being dumped.

We asked, "Have you ever heard of a UFO with Blockbuster ads hanging on it?"

"Absolutely," he mumbled through the gravel. "These ads are camouflage to hide the real identity of these invaders. They lull us into a false sense of security. They throw us off our guard so the aliens can land without being detected."

This agent worked directly under CIA head Bill Casey during Irangate. Blockbuster vehemently denies advertising on UFO's.

"Good Gawd!" We were aghast. We hadn't dreamed the UFO's had landed. This fact was not reported by Gulf Breeze.

The thought of alien space creatures establishing a beachhead and wandering loose among us is scary. We checked various bars in Pensacola to be sure the invaders had not snuck across the bay bridge and invaded downtown, lurking undetected among our innocent citizens.

We questioned Wilmer Mitchell, chief innkeeper at Rosie O'Grady's. "Come to think of it, I did notice some strange looking patrons hanging around the End of the Alley Bar last night."

Beware. Be on the lookout. UFO's are confirmed; they may have actually landed. We do not know whether these aliens are friendly or not.

Don't be fooled by negative Pensacolians who are jealous just because the UFO's prefer to land in Gulf Breeze. The notion that what they saw was a blimp is just sour grapes.

If you should happen to meet a mysterious character in a bar, be on the alert. He may be from outer space. Report all such incidents to the Gulf Breeze Police Department.

We can't be too vigilant. One of these alien characters might try to run for public office.

Pass the Broccoli ... Again

> I sure hope this rocket wasn't assembled by the lowest bidder.
> *Astronaut*

As we sit here looking out into space, we are trying to figure out what a trip to Mars will really be like.

The U.S. has already been to the moon, but Mars is a far piece beyond that, 2000 times farther, to be exact. It's so far away that it will take a year for a space ship to coast up there. It will take 41 minutes for a phone call to get there.

We can just see the two astronauts on their way to Mars. We think the trip will go something like this:

Astronaut Number 1, yawning: "What are you doing floating around over there, Number 2?"

Number 2: "I just called my wife. If she calls right back, I should have an answer in 82 minutes. I can't remember the kids' names."

"Too bad. What are you reading?"

"I'm reading *War & Peace* for the 25th time. It gets worse and worse. How about you?"

"I'm reading the phone book. The plot's no good, but it has a great bunch of characters."

"Is it my turn to go to the bathroom?"

"Yeah, according to the calendar. Don't lose my place in the Sears catalog. There's a great story on page 479. And leave the seat up."

"The seat doesn't work. It's that damned $26 million commode that NASA designed. It's spring-loaded down. It's dangerous. I got trapped in it yesterday."

"Incidentally, where is the potty? I forgot to put the anchor down."

"Last time I saw it, it was headed for the kitchen."

"Wonder what's going on down on earth?"

"Just the election. I'm glad our TV is busted. But I've already forgotten who's president. I think it's still Nixon."

"My God, look out the window. I think we just passed my mother-in-law. I didn't think she was up here."

"I thought I saw mine yesterday, but that's a mistake. Mine went the other way."

"I haven't heard from my wife yet. This sure is a good way to get out of the house for a couple of years, but don't tell my wife I said so."

"What's your wife's name?"

"I think it's Helen."

"What's the score in our card game?"

"You owe me $473,000. We'll play some more after dinner. Whose turn is it to cook?"

"Mine, I guess. Boy, I sure could use a martini. How about opening the bar early tonight? I feel like Happy Hour."

"OK. What's on the menu? I'm sick of that dehydrated hamburger extender Wellington you keep dishing up."

"Don't knock it. The extender is all gone. We're down to broccoli."

"Good Lord, you mean we have another 163 days of that stuff?"

"Afraid so. I hope the food is better on Mars. What do you suppose those people eat?"

"That's what they sent us up here to find out. I hope to hell they don't grow broccoli."

"I sure hope we find somebody up there to talk to. I'm sick of listening to you."

"Same here. What do you think Martians look like?"

"The people in Gulf Breeze think they look just like the people in Gulf Breeze. That rock with the worm in it sure didn't look very promising. Maybe we can bring some aliens back with us. I hope we find more than angleworms up there."

"What if they take us prisoner?"

"Maybe they'll fly us home in their UFO. The people in Gulf Breeze would love that. We can land right on City Hall."

And so it goes for 365 days one way, and 365 days back. Perish the thought!

Horse Thieves

It is ridiculous to spend $250 looking up your family tree. Just try scraping somebody's fender, and it's amazing what he'll tell you about your ancestors.
Mickey Marvin

Beware of family reunions. You have to go because at our age you never know when it will be your last one, and you might miss some good gossip.

We recently attended one such event in San Diego. The host has a beautiful home perched high on a hill overlooking a valley, a lake, the mountains, and an interstate with California's bumper-to-bumper traffic.

Southern California has ideal weather if you stay out of the desert. Dry, mild, clean air, lots of sun, little rain, no humidity. Annual rainfall is less than the ten inches which Hurricane Danny dumped on us here. Compared to California, Florida feels like a sauna.

The last time we attended one of these affairs we naturally prepared to stay a decent length of time until the host cut off the water. We drank out of a hose and did what we had to do behind a big bush.

To save money on this trip we had planned to take a migrant bus and pick cotton, fruit, and vegetables on the way, but we missed the bus and had to fly Budget Airlines.

For our visit the host installed a temporary outhouse in the front yard. This was no idle threat considering his previous reputation for treachery. We remained under the constant threat that he would cut the water pipes again and force us into the front yard.

Prior to the visit we received an elaborate military field order which laid out a comprehensive agenda for the reunion. The host carefully detailed what he would pay for and mostly what he wouldn't. The guest-pay items were listed as "no-host."

We didn't understand this no-host wrinkle since he was the host, and every reunion needs a host. How can you have a no-host host? When questioned about this discrepancy, he said, "Bring your wallet." A subtle hint. Another relative joined us in voting against this no-host scam. We were outvoted one to two. Democracy did not prevail.

We did try to pay for one expensive family dinner, but we forgot our wallet. We were embarrassed but relieved. We did volunteer to pay for a cheap lunch the next day.

The agenda included some menu items which the hostess provided. One item was salmon *à la dishwasher*. We got a big chuckle out of this funny little joke. A host with a real sense of humor. Then we arrived. The hostess served salmon piping hot straight out of the dishwasher. This was no joke.

It seems that you can poach salmon by running it through a complete dishwasher cycle. It was honestly the best salmon we ever tasted. We are going to try this ourselves using the washing machine, dryer, and heat pump. We have learned you do not put the salmon in with the dirty dishes or laundry. It goes through solo.

Another menu item was listed as *al fresco*. We had never tasted *al fresco* before. It turned out *al fresco* means out-of-doors, and we were served outside. We had to eat in a hurry because the weather got cold. We then rushed to bed before the host turned off the electricity. He handed out blankets in the dark.

The host arranged golf—no-host, of course. He did furnish clubs, golf balls, tees, water, soda, and snacks. The clubs were nice but did not work very well. We learned later he kept a special set of clubs and some secret golf balls with which he thoroughly trounced us all. We had to pay for this humiliation.

One relative lost all his baggage. The host blamed the airline, but we discovered later that he hid the bags for fear we wouldn't go home.

All in all it was a grand visit, host or no-host. The hostess was nice, and we took group pictures sitting on a rock. We told lies and jokes and tried to reconstruct our family tree. We traced the branches into various cemeteries. One earlier ancestor had traced us back to a gang of horse thieves so we didn't go back any further for fear of what might turn up.

Finally the host cut off all the power and water ... permanently. We were forced to go *al fresco*. Then he padlocked the outhouse, and we came home, no-host. The host and hostess personally took us all to the airport to make damned sure we left.

On the Road Again

> Thanks to the interstate highway system, it is now possible to travel across
> the country from coast to coast without seeing anything.
> *Charles Kuralt*

Travel is good for the soul so we loaded up the family and drove to Washington, D.C., and points along the way.

Charles Kuralt is right about the Interstates. In fear of your life, you can't look right or left, just straight ahead. You drive 70 m.p.h., and you're standing still. The traffic roars by you at somewhere between 85 and 150 m.p.h., including hordes of huge trac-tor-trailers. The public ignores the speed limits. You just swallow their exhaust. The Interstate is virtually a raceway. As Mark Twain explained, "A difference of opinions is what makes auto racing."

Traffic does move briskly until there's a wreck, then absolute gridlock. Traffic stops and backs up for miles. Road rage sets in; the language is not pretty.

In order to see the world, you have to get off the Interstate. We made one pilgrim-age to Williamsburg, VA. This handsome colonial village takes you back 200 years to the simple life. No indoor plumbing, just bed pans; no cars, just horses and carts; and lots of taverns with lots of ale and public and private accommodations.

The taverns stacked their sleeping guests in like cordwood, two to a bed. Lots of strange bedfellows. It was safer to travel with a bedfellow you knew.

Now rebuilt by the Rockefellers, the town is authentic, manicured, and immacu-late, tourists and all. This is our kind of town—we were born 200 years too late.

Moving right along, we also slipped off the Interstate raceway at Chapel Hill, NC, home of the University of North Carolina and the "tarheels." Chapel Hill is a perfect college town with the neat, clean, well-manicured college campus and magnificent trees. The town is the dogwood capital of the world; the storm of white blooms there in the spring looks like a Minnesota blizzard.

Actor Robin Williams just happened to be on campus making a movie about "Patch Adams," a real and renowned character who also happened to be a renowned pediatrician. Patch had a way with children.

It seems that when Patch graduated from college, he mounted the platform, re-ceived his diploma, and paraded out of the hall with absolutely no back in his gown, no pants, no drawers, no nothing. Completely bare! We got to see Robin Williams play the part to perfection. We had seen Robin from the front before, but we had never seen his rear end. Katharine Hepburn once allowed that "Acting is a minor gift . . . . After all, Shirley Temple could do it at age 4."

Then on to Washington, D.C., the nation's capital, to check on the latest scandals. Washington is sometimes referred to as Foggy Bottom, for good reason.

Foggy Bottom

> A man climbed over the fence, ran across the White House lawn, and pounded
> on the door, yelling, 'I gotta see the president'.
> Vice President Quayle shouldn't have to do that.
> *Bob Orben*

Entering Washington, D.C., by car you drive through sadly neglected and dilapi-dated neighborhoods, then suddenly enter the prestigious hub of our nation's capital: the Capitol, grand buildings, the White House, proud monuments, and scads of bu-reaucrats and tourists. With all of the cabinets and bureaus and departments, it's no wonder the government doesn't work.

Our very first visit in the capital was to the White House to check on the latest skulduggery. Night fell. Suddenly everybody else was gone, and we found ourselves locked inside the White House grounds, ringed in by the Secret Service. Through the bars we begged to get out. No use! We were hemmed up like Al Capone.

There were hushed rumors of a big event coming. Could it be Monica? Now we refused to leave for fear of missing any excitement.

Sure enough. We heard a helicopter chugging in to land. Minutes later an impressive motorcade wheeled onto the grounds: police cars, two limousines flying flags, six or eight vans, and an ambulance. Here came the president himself and the press riding in from the helicopter several blocks away. A golf cart would have been more economical, but this is the federal government at work, remember.

The Secret Service explained that when Hillary was asleep in the White House, the president didn't like to land on the south lawn and wake her up. A likely story.

We later learned that the Secret Service had let all Democrats out of the gate but had only locked up Republicans for fear we would get out and vote. We were finally released after we promised to vote Democratic.

The next day we were let back in to see the inside of the White House. It is certainly a fitting and impressive residence for our presidents, grandly decorated with richly antique appointments and filled with historic portraits save one, a Georgia O'Keefe rendering somebody snuck in there lately. It looked like the bastard at the family reunion.

Great tales abound about our past presidents. William Howard Taft, later chief justice, stood 6 feet 2 inches tall and weighed in at 330 pounds. He got stuck in the White House bathtub, and the fire department had to come and pry him loose with a hoist.

President William Henry Harrison spoke for 1 hour and 40 minutes at his inauguration in the cold rain with no coat, no hat, and no umbrella, the longest inaugural speech on record. He died 30 days later of pneumonia. His grandson was elected president in later years and delivered the shortest inaugural. Small wonder.

After driving 2500 miles on our trip, the water pump blew up 120 miles from home. Five of us came home in the front seat of the wrecker towing the car behind us at 75 m.p.h. with a lady wrecker driver. A female Barney Oldfield.

Travel is good for the soul but not for your wallet or your vocabulary. Travel is better at home.

ON YOUTH

Grandchildren and elephants never forget.
Unknown

I learned that my daddy can say a lot of words that I can't.
Unknown 8-year old

We went to a tough high school. In creative writing we learned to
write effective ransom notes.
Dan O'Day

If you think education is expensive, try ignorance.
Bumper sticker

If it was going to be easy to have kids, it never would have started
with something called labor.
Anonymous

Clancy Lowers the Boom

Grandchildren are your reward from God for not killing your children.
Author unknown

We remember a challenging and exhilarating babysitting assignment with our favorite granddaughter. She was only six at the time. Her name is Clancy, all Irish, with everything the name implies.

There was good news and bad news. The good news was that we arrived safely. The other news was much more interesting.

Clancy's mother, our favorite and only daughter, carefully prescribed our babysitting duties in writing. Everything was legal.

Our first major duty was to administer antibiotic medicine to Clancy and heartworm medicine to the dog. The dog is very pedigreed and very wild when she and Clancy collide in the living room. We went the full ten rounds with Clancy and the dog and the medicine. In the end, after the lights went out, we think Clancy got the heartworm medicine. The dog is still all right, and Clancy showed no signs of heart-worms or other ill effects.

Next on the list we were directed to dress Clancy in her finest and escort her willingly or otherwise out to dinner. Dinner went well, including a large $25 steak which she devoured. Clancy likes big steaks. Just as we were congratulating ourselves on a great success, Clancy disappeared. We were frantic. A search party set out immediately.

We found Clancy at the front door holding a large crowd at bay in the parking lot where she was hurling rocks. Prospective dinner guests were unable to get into the building for fear of being struck by the flying missiles. Clancy throws a great fastball. Neither the crowd nor the manager was amused at her athletic prowess. We managed to relieve her from the mound just as she retired the side and the crowd turned ugly and threatened to storm the building. No one was seriously injured.

Last on the prescribed duty roster was a visit to the department store to buy Clancy an Easter dress. This turned out to be a fairly serious mistake.

While we were attempting to warn the children's department manager what might be in store, Clancy disappeared again. A thorough search began, and an emergency missing persons bulletin was broadcast throughout the store. Alarms, bells, and whistles went off. By the time we found her, it was too late; the damage had already been done.

Clancy systematically rearranged all the children's clothing, price tags, and sizes according to her own system. It would have been much simpler to unscramble eggs than to sort out and rearrange the merchandise. The department manager was not amused. Neither were we. Clancy needed a remedial application on her posterior.

We are back home now. Clancy is forgiven, although her mother has not to date forgiven us; but Clancy has forgiven us and has already invited us to come back.

Somewhere there must be a law against abusing grandparents. We need to organize for protection. Clancy has already organized the grandchildren.

More Kids

> Youth is such a wonderful thing. What a shame to waste it on children.
> *George Bernard Shaw*

There's bad news and good news.

Our teenagers lead the world in drugs, guns, shootings, murders, and pregnancy. However, U.S. Attorney General Janet Reno just announced that teenage crime is down 9.2% so the news isn't all bad.

The truth is most of our children are great kids so let's hear some stories about what they're up to these days:

A young man who had just started to drive proudly announced to his father, "Dad, I just backed the car out of the garage."

"Son, that's wonderful. I'm proud of you."

Said the son, "The trouble is I backed it in first."

In English class the teacher asked the children to write a piece of fiction which included elements of deity, royalty, sex, and suspense. Little Jimmy finished in less than three minutes. The teacher was amazed.

"Jimmy, how could you possibly write fiction in such a short time. Please read it to the class."

Jimmy read, "'My God,' said the princess, 'pregnant again. I wonder who done it.'"

During the last census a census taker knocked on the door of a residence as part of his duty to record the necessary census information. A young boy answered the door.

"Son, is your father home?"

"No, sir, he's in jail."

"Is your mother here?"

"No, sir, she's in the insane asylum."

"Well, is your brother home?"

"I got two brothers. One's in reform school, and the other one's at Harvard."

"Harvard University. My, that's wonderful. What's he studying?"

"He ain't studying. They's studying him." So much for counting people.

In another family the older brothers broke a window. The little brother wanted to tell on them to their parents. The older brothers tried to bribe the youngster not to tell.

"How about if we each give you a dollar?"
"I don't want any dollars."
"Well, how about if we buy you some ice cream?"
"I'm not hungry."
"Can't we take you to a movie?"
"I don't wanta see a movie."
"Well, what the heck do you want?"
"I wanta tell!" You can't win them all.

A relative was introducing a child to the rest of the family at a family reunion.
"Who's that?" asked the boy.
"That's your grandfather on your mother's side."
"Now who's that over there?"
"That's your grandfather on your father's side."
"Well, if he ever wants to get anyplace, he had better get on my mother's side."

Out of the mouths of babes come the very choicest comments. Parents love their children, but children can also drive them half-crazy at times.

A wise man said, "Adolescence is a period of rapid change. Between the ages of twelve and seventeen, parents age twenty years."

Little League Supporters

> Rich Rolkers is throwing up in the bullpen.
> *Jerry Coleman, baseball announcer*

A serious problem has arisen in Boca Raton, Florida, where an umpire removed one Miss Melissa Raglin as catcher of her youth league Dodgers and sent her to the outfield. Why? Because Melissa wasn't wearing a cup as decreed by the league rules for catchers.

Never mind that she was a team star and had already been catching for the Dodgers for two years without a cup. Can't you just see the scene when this unfortunate incident occurred:

Umpire: Are you wearing a cup?
Melissa: What for?
Umpire: For catchers. (He hands her an athletic supporter complete with cup.)
Melissa: What's this?
Umpire: It's a jock strap.
Melissa: (trying to fit the cup over her nose): This thing won't work. It doesn't fit. I can't see through the straps, and it won't fit under my mask. Anyway, my nose is OK.
Umpire: No, no, you don't understand. That's not a nose guard. Here, let the pitcher explain it.
Pitcher: (embarrassed): This is for what I've got that you don't have.
Melissa: I don't have one?
Pitcher: I don't think so.
Melissa's Mother (irate): This is ridiculous. Ump, can't you understand? Melissa's a girl. She doesn't need that contraption. She just needs to catch.
Umpire: I don't care what you think. It says right here on page 472, Rule Number 7,864, that all catchers will wear cups, no ifs, ands, ors, or buts. It don't say nothing about girls. It just says catchers. Melissa's a catcher.
Melissa: I won't wear that silly thing. I'm a girl.

Umpire: Well, you're outta here. Get out in right field. You don't need a cup out there.

The next day Melissa arrives ready to catch again.

Umpire: Where's your cup?

Melissa: Right here on my ankle.

Umpire: That ain't for your ankle. It ain't gonna do you any good down there. Outta here!

Finally the team comes up with a ladies jock strap used by female Karate contestants. Melissa is now wearing this garment and catching again where she belongs. We have not heard the last of this matter, however; Melissa's mother sues. Can you imagine the U.S. Supreme Court batting this matter around among seven male justices and two female justices?

Justice Sandra Day O'Connor: Here's the evidence. (She hands the jock strap with cup to Justice Ruth Bader Ginsburg.)

Justice Ginsburg: What is this?

Justice O'Connor: It looks like a nose guard.

Justice Ginsburg: Well, it's too darn big, it doesn't fit, and I can't see through these darn straps.

Justice O'Connor: Well, maybe it's for your ankle. Try that.

Justice Ginsburg: There, that's better. But I'll need two of them.

Chief Justice William H. Rehnquist turns red, white, and blue and all other colors of the rainbow.

Chief Justice Rehnquist: No, no, ladies, you've got it all wrong. That's to protect the family jewels.

Justice O'Connor: But I don't have any.

Chief Justice Rehnquist: Oh, lordy, that's right. I forgot.

The chief justice calls a quick recess, and they retire to chambers where a male justice models the jock strap. Then they return to the bench.

The vote is 7-2 for jock straps. Male chauvinism is not dead yet.

Babies Having Babies

An ounce of prevention is worth a ton of cure.
Anonymous

Illegitimacy is in!

One-third of our babies born are bastards.

It is now very fashionable to sleep in and have babies without benefit of matrimony, Hollywood-style. Half of our children are being raised in single parent families, mostly mamas with AWOL papas.

How do we stop this flood of illegitimate babies, mostly mothered by teenagers, which is the biggest tragedy of all? Many of these kids graduate into permanent poverty.

Now comes an interesting solution: "baby" dolls. An enterprising couple has invented a computerized, life-like doll which does all the things that real babies do.

If a teenager has a glint in his or her eye indicating a propensity for promiscuity, he or she can get one of these dolls on a trial basis before the real thing happens.

We wondered how this preventive measure might work out. First we visited a young high school girl who had shown signs of straying from the straight and narrow. She was mothering a baby doll.

"How's the baby doll working out?" we asked.

"Right now it's asleep, but it's a constant battle." Just then came the sound of a screeching infant.

"There it goes again, just like an alarm clock, every four hours. It's hungry, and I have to get it up or it won't stop that infernal racket."

She retrieved the doll baby, fed it some leftover spaghetti, then put it back to bed. No sooner had we started to talk again than the doll hollered again.

"! 0 # * ! ? ! There it goes again. Now it's wet. Diaper time. I can't stand any more of this crap. No rest, no sleep, no let-up, no fun, no nothing. Just bottles and poop!" She picked up the baby doll and threw it out the window.

I'm cured," she said. The glint in her eye was gone. So much for this frustrated, would-have-been mother.

Then we wondered how a baby doll would work on a young man with the urge. We went to see one who was fathering a doll.

"How are you making out with your baby doll?" we wanted to know. He was rocking and burping his doll.

"# 0 * ! ? * ! I didn't know it would be nuthin' like this. It's hell after the fun is all over. I didn't count on all this screaming and burping and dirty pants. It's a pain in the you-know-what!"

"Have you learned your lesson?" we wondered.

"I sure have," he admitted. "I didn't know nuthin' like this would happen. All I do is burp this damned doll and change its pants."

He took the baby and dropped it down the elevator shaft along with the dirty diapers. The urge had departed him; the glint was gone.

"That's the end of that. I'm cured," he smiled.

So goes the experiment. This idea may just work out after all. Fewer babies having babies.

This is real character building. Somebody wisely said, "Character is what happens when the lights go out."

Please Pass The Ketchup

> They knew no men could ever be so hungry that they would eat a lawyer.
> *Will Rogers on Cannibalism**

If you thought cannibalism was dead, think again. There were times in olden days when tribes of cannibals did roam about dining on visitors.

Mark Twain once reminded us of history, noting that "The natives finally understood the missionaries. They ate them." But Twain had a reputation for hyperbole.

Will Rogers, however, did recount one true story about a Harvard lawyer named Packer in Colorado who ate his companions on a camping trip. This incident happened to be the only reported case of cannibalism at the time.

A Republican judge sentenced Packer to hang since he had eaten all the Democrats in Hillsdale County, but on appeal Packer got 40 years instead, which was a little over 6 years for every camper he ate. "You had to eat 10 or more to get life," reported Rogers.

Well, Packer didn't even serve that long. There happened to be a close election coming up so the Republicans let him out early, hoping he would eat more Democrats before the vote on November 4.*

The days of cannibalism are now long past, or so we had hoped. This is the enlightened age. But, believe it or not, we now have reports of more cannibalism in 1997. Mattel has manufactured and sold a Cabbage Patch Kids Snacktime Doll which eats small children.

It seems that this doll is only supposed to eat plastic french fries and carrots which then exit and are deposited in a pouch on the doll's back. Mattel did have the good grace not to excrete the fries and carrots in the normal manner.

A plastic snack is not very appetizing so the dolls started eating parts of their little owners. Little girls lost fingers, hair, and scalps inside these monsters.

While trying to save one of his little customers, a barber completely dismembered the doll in order to stop it before it could eat both him and the little girl. It seems that Mattel forgot to install an on-off switch although the simplest, quickest way to stop the doll is to jerk out the batteries which power the jaws. Who might next fall prey to these cannibals? Mothers, brothers, grandparents, anyone. We have a serious health hazard here.

Enter a lawyer, possibly from Harvard, on behalf of one little victim. He is suing Mattel for 25.5 million dollars, claiming the whole family will require therapy because of the attack on their daughter. "Mental and emotional injuries," he claims.

25.5 million dollars seems a fair figure. That's $50 in actual damage rounded off to the nearest $25.5 million. The lawyer said he got this figure by estimating Mattel's total profit on these person-eating dolls and then multiplying that by three; he would get only 1/3 of that for himself, however. That certainly makes sense.

So far there have been no reported attacks on lawyers. It would be very difficult to digest an attorney.

*From *Will Rogers: Reflections and Observations* by Bryan B. and Frances N. Sterling, Crown Trade Paperbacks, 1986.

Teens

> Insanity is hereditary. We get it from our children.
> *Unknown*

The bad news is that U.S. teens lead the industrialized world in guns, murders, pregnancies, drugs, suicides, gangs, and poor math and science scores. The good news is that most of our teens are good kids who stay sane, healthy, and out of jail, and study if required to do so.

We don't hear the good side often enough. Our good teenagers even have a good sense of humor, and there are a lot of great stories about some of their doings.

For instance, President Felton Harrison of PJC used to tell the story about a teen-age college student who was hitch-hiking back to school. A farmer with six hogs in the back of his truck picked him up and let him ride in the back with the pigs.

It just so happened that this occurred during the Mediterranean fruit fly invasion in Florida, and the department of agriculture was stopping all vehicles at each county line to inspect for fruit flies.

"What do you have in the back of the truck?" asked the officer.

"Six hogs and a college student," said the farmer and drove on to the next county line.

"What do you have in back?" asked the officer.

"Six hogs and a college student," replied the farmer.

After the same thing happened again at the next county line, the student finally spoke up.

"Sir, may I ask a favor?"

"Sure, what is it?" asked the farmer.

"Sir, at the next stop would you mind introducing me first?"

High school can be just as much fun, but some high schools can be tough. Dan O'Day reports he knows of one high school which is so tough that "in creative writing the students learned to write effective ransom notes."

But they're not all that tough although getting good grades can be a problem. One young man brought home a poor report card and confronted his father. He handed over his report card.

"Dad, here's mine and one of yours I found in the attic."

Another young man was not so lucky. He brought home four F's and a D.

"How did you get this D?" asked the father.

"Well, dad, I'm sorry. I spent too much time studying the other four subjects."

Unfortunately some students have a habit of copying other students during tests. During one exam the copier sneaked a look at his classmate's paper and observed that in answer to question number 16 his classmate had written, "I'm sorry, but I don't know the answer to this question."

On question 16 on his test the copier wrote, "Me neither."

Kids Will Be Kids

> By the time a man realizes that maybe his father was right,
> he has a son who thinks he's wrong.
> *Charles Wadsworth*

You never know what will happen to your children next. Very recently a very young man in early grade school made the fatal mistake of giving a lemon drop to a little girl classmate. Boom! Chivalry is dead, dead, dead.

The school authorities descended on this poor young criminal for dispensing an unauthorized, unapproved, unidentified substance to a minor. He was reprimanded and suspended for half a day. Now the miscreant has a criminal record and is identified disparagingly as the "Lemon Drop Kid."

This event would be funny if it were not so tragic. Somehow the whole world is paranoid. But kids are funny, and we should appreciate their humor.

Dan O'Day tells this story:

For school lunch one day the cook served venison. She asked the kids to guess what it was. "I'll give you a hint, she said. "It's what your mother sometimes calls your father."

One little boy stood up and screamed, "Don't eat it! It's jackass!"

Another little boy got lost at the zoo. He went up to a guard and asked, "Did you see a lady pass by here without me?"

A not-so-little girl asked her mother, "Mama, do you remember what we saw the animals doing at the zoo when I was four years old?"

"Yes, dear. What about it?"

"Well, I'm pregnant."

When a little boy came home from Sunday school, his mother asked him about the lesson. Wide-eyed, the little boy explained the parting of the Red Sea. "They brought in a construction company with demolition engineers and marine engineers and all sorts of heavy equipment."

"But dear," said the mother, "is that the way the Bible tells it?"

"Mama, if I told it the way they tell it, you'd never believe it."

One cold day, a farmer brought some eggs into the warm kitchen and let the cat sit on them. When his son came in and saw what was happening, he was visibly astonished and mortified.

"What's the matter, son?" asked the father.

"I'm sorry, dad, but I've done 'et my last egg."

Before a little boy went off to a dance, his mother warned him, "I want you to be nice to the little girls and be sure to say something nice to everybody you dance with."

Sure enough, during his very first dance, he said to his partner, "You sweat less than any fat girl I've ever danced with."

You can learn the great lessons of life from your children:

Fran Lebowitz tells us, "Never ask your child what he wants for dinner unless he's buying."

The easy way to teach your children the value of money is to borrow from them.

One seven-year-old quickly discovered that you can't hide your broccoli in a glass of milk.

One particularly bad little boy prayed, "Try hard. If at first you don't succeed, try, try, try again."

Enjoy your children and grandchildren. They won't be young forever.